Evoked Brain
Potentials in Psychiatry

Evoked Brain Potentials in Psychiatry

Charles Shagass, M. D.

Professor of Psychiatry
Temple University Medical Center and
Chief of Temple Clinical Services
Eastern Pennsylvania Psychiatric Institute
Philadelphia, Pennsylvania

 PLENUM PRESS • NEW YORK–LONDON • 1972

Library of Congress Catalog Card Number 76-157928

ISBN 0-306-30533-X

© 1972 Plenum Press, New York
A Division of Plenum Publishing Corporation
227 West 17th Street, New York, N.Y. 10011

United Kingdom edition published by Plenum Press, London
A Division of Plenum Publishing Company, Ltd.
Davis House (4th Floor), 8 Scrubs Lane, Harlesden, NW10 6 SE,
London, England

Printed in the United States of America

Preface

Two purposes have guided the writing of this book. Originally, I wanted only to bring together the results which we have obtained during more than ten years of work on evoked potentials in psychiatric disorders. However, it soon became clear that I really wanted to do a little more than that. First of all, a systematic review of the literature seemed indicated. Even though research findings are usually presented in the context of such a review, our laboratory has not studied every aspect of evoked potentials. Consequently, it seemed more appropriate to place our own results within the framework of a general presentation of the evoked-potential field, rather than to have our specific studies govern topic selection. Second, I found that I wanted to expound on the principles and details of techniques to a broader extent than warranted for presenting only our own results. The motivation for attempting such a "methodological primer" came not only from my long-term preoccupation with technical issues, but from contacts with many investigators who consulted me during the early stages of their ventures into evoked-potential research. Thus, to the initial goal of a research monograph was added that of a systematic account of both the substantive findings and the methodology of the field.

I owe thanks to many colleagues and to my staff. Although I must take full responsibility for the contents, I should like to acknowledge the substantial contributions of Donald A. Overton to the chapter on methods, both in stimulating my thinking and in critical review. The contributions of my other collaborators are documented in the literature citations. A large number of technical assistants helped to gather our data; among these, Dewey M. Trusty deserves special mention for the diligent application of his talents that permitted us to obtain acceptable recordings from disturbed subjects under the most trying conditions. Our work would not have been possible without the instruments provided by Harold W. Shipton and John W. Emde and, more recently, by Stephen Slepner.

Also essential to the conduct of the research was the financial support given by the National Institute of Mental Health, United States Public Health Service, through grants MH 02635 and MH 12507. These grants supplemented the splendid basic support provided by the Departments of

v

Psychiatry at the University of Iowa and Temple University and by the Eastern Pennsylvania Psychiatric Institute. Thanks are also in order for the assistance provided by a grant from the Iowa Mental Health Research Fund.

The onerous task of typing and retyping the manuscript was accomplished by Regina McHenry and Ann McGrath, to both of whom I extend sincere thanks.

No list of acknowledgements would be complete without mentioning Herbert H. Jasper, my teacher in electroencephalography, who first suggested to me, in 1954, that evoked potentials should be used to study cortical excitability in psychiatric disorders.

Contents

Chapter 1

Introduction

The development of methods for recording evoked potentials in man presented psychiatric research with an exciting opportunity. The brain's electrical responses to sensory information could now be registered in the intact human being, capable of following instructions and reporting his experiences. Responding early to the opportunity and challenge offered by the method, we have been exploring the psychiatric correlates of evoked responses in our laboratory for more than a decade. This book contains an account of what we and other investigators have so far found.

To date, only a small portion of the human evoked-response literature deals specifically with psychiatric problems. Much more work has been done in closely related areas of psychological research; for example, many studies have dealt with the phenomena of attention. Such psychiatrically relevant topics have been included here. It has also seemed desirable to consider methodology in some detail. Although the problems of psychiatric research have been given special attention, it is hoped that the discussion of methodological issues will also be applicable to evoked-response work in other disciplines.

SOME HISTORICAL BACKGROUND

Richard Caton (1875) was the first to demonstrate that the brain responds electrically to stimulation of a sense organ. He put one recording electrode on the exposed cerebral cortex of an animal and the other on a cut surface and, using the light from a lamp as a stimulus, observed changes in potential. Not only was Caton the pioneer investigator of sensory-evoked potentials, but he also discovered the electroencephalogram (EEG) in in the same experiments. With both electrodes on uncut brain, he found incessant oscillations of current in the absence of stimulation; this was the

1

EEG. It was about a half century from Caton's first report to Hans Berger's (1929) discovery that the EEG can be recorded through the unopened skull of man. Following Adrian and Matthews' demonstration in 1935 that Berger's rhythms really originated in the brain, electroencephalography developed rapidly. This development was aided by the growth of electronics, by early demonstrations that there were distinctive EEG waveforms in epilepsy (Gibbs *et al.*, 1935), and that the EEG could help to localize brain lesions (Walter, 1936).

The sensory-evoked potentials were not, however, so easily recorded from the human scalp as the EEG. They are more attenuated in being conducted through extracerebral tissues and, under ordinary circumstances, are almost totally obscured by the larger "spontaneous" rhythms. Thus, although evoked potentials provided a useful research tool for neurophysiologists who could place electrodes directly on or into the brain substance, it was necessary to devise special techniques to record them in the intact human.

George D. Dawson (1947) was the first to report a method for detecting cerebral action potentials evoked by peripheral nerve stimulation from scalp recordings in man. His method was based on the assumption that the evoked responses would occur at a fixed time after application of the stimulus, and be relatively uniform in wave shape, while the timing of the "spontaneous" brain rhythms would be random with respect to the stimulus. Following nerve stimulation, Dawson displayed EEG traces on a cathode-ray oscilloscope and superimposed a number of these traces on a single photographic record. He was then able to show that certain EEG events do take place in consistent time relationship to the stimulus and that these time-coherent events, the evoked potential, can be distinguished from the background rhythms. Figure 1 illustrates Dawson's superimposition technique.

Dawson (1951, 1954) subsequently devised an instrument for automatic summation to detect small evoked potentials. Although based on essentially the same principle as the superimposition method, namely, that time-locked activity would retain its form in summation and random activity would

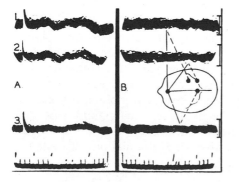

Fig. 1. *A*, 50 superimposed cathode-ray-oscilloscope traces containing responses to stimulation of the left ulnar nerve at the wrist. The stimulus escape at left is followed by a train of waves, each with a period of about 20 msec. *B*, similar recordings without application of a stimulus; no deflections comparable with those in *A*. Time scales show intervals of 5 and 20 msec. Calibrations at right equal 20 μv. (Reprinted with permission from Dawson, 1947.)

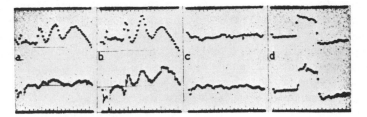

Fig. 2. Cerebral responses to electrical stimulation of left ulnar nerve at wrist. In *a,* upper trace is from electrodes on the right side of the head and lower trace from the left side. Stimulus applied 6 msec after start of sweep. In *b,* both traces recorded from right side of the head; upper trace over sensory area, lower trace 3 cm behind it. In *c* are control records with no stimulus applied. Average of 220 sweeps in all cases. In *d,* calibration pulses of 2.5 μv. Positivity in presumably active electrode gives upward deflection. (Reprinted with permission from Dawson, 1954.)

sum to a horizontal line, the summation procedure provided clearer results (Figure 2). Dawson pointed out that, in other disciplines, averaging had long been applied to the detection of systematic fluctuations amongst larger, irregular ones. For example, Laplace had predicted, in the eighteenth century, that by averaging enough data, it should be possible to demonstrate a lunar tide in the atmospheric pressure, and this prediction was verified in 1847. Radar provides a more recent application of the averaging principle.

Following Dawson's lead, a number of evoked-response summation or averaging instruments were devised during the 1950's. However, widespread utilization of evoked-response recording methods in man occurred only after introduction of commercially manufactured special-purpose digital computers for averaging (Clynes, 1962). Such computers found quick acceptance in many laboratories and the small amount of work on human evoked potentials existing before 1960 was augmented with such rapidity that a symposium held in February of 1963 resulted in a volume of over 500 pages (Katzman, 1964*a*). The growth of the literature on human evoked responses has continued at an ever increasing rate, and specifically psychiatric studies have contributed to this growth.

DEVELOPMENT OF THE WRITER'S EVOKED-RESPONSE RESEARCH

Our interest in the use of evoked-response methods for psychiatric research began with studies of EEG responses to intermittent photic stimulation. Brief flashes of light presented to closed lids at rates in the vicinity of the alpha rhythm frequency (10 Hz) will generally elicit rhythmic activity at the flash rate. This response is called photic driving. The amount of photic driving response varies considerably from person to person, and we attempted to relate these variations to the affective state of psychiatric patients. In one study, the relative amount of driving to flash rates of 10 and 15 per second provided significant differences between patients who were predominantly depressed and those who were mainly anxious (Shagass, 1955a). The patients with anxiety had relatively more driving at the faster flash rate, whereas those with depression showed more driving at the slower flash rate. Nonpatients gave results that were intermediate between those of the depressed and anxious groups. Longitudinal studies of a few subjects also showed that day-to-day fluctuations in affective state were correlated with photic driving (Shagass, 1955b). However, our attempts to repeat these observations yielded variable results. Ulett et al. (1953), who had found significant relationships between photic driving and anxiety-proneness, also encountered difficulty in replicating their results (personal communication).

The disappointing outcome of the photic driving studies led us to search for indicators of cerebral responsiveness that might be more stable. The report by Gastaut et al. (1951a) on measurement of the cortical excitability cycle suggested interesting possibilities. The method employed by these workers involved stimulating the subject with two light flashes. The interval between flashes was varied from trial to trial. They found that the response to the second (test) flash varied in amplitude as a function of the time between the two flashes. The relative size of the second response, compared to that of the response generated by the first (conditioning) flash, was taken as the index of excitability. The time course of such recovery could then be graphed. In animals, Gastaut et al. (1951a) demonstrated phases of the excitability cycle that were analogous to those shown in classical studies of peripheral nerve, i.e., periods of absolute and relative refractoriness and phases of subnormality and supernormality (Figure 3). They showed that the recovery cycle was modified by drugs affecting the central nervous system and that the nature of these modifications was in a predictable direction. For example, recovery was facilitated by a central excitant like pentylenetetrazol (Metrazol®) and retarded by depressants, such as barbiturates. They were also able to record and measure recovery cycles in human subjects with unusually large responses without the use of averaging (Gastaut et al., 1951b).

Gastaut's findings suggested that our photic driving results in anxiety and depression could be a manifestation of different cortical excitability cycles in these affective states. During 1954, we attempted to test this hypothesis by using Gastaut's method in patients. However, without an averaging instrument it was not possible to measure responses with any degree of confidence and the attempt was abandoned. In 1956 we decided to pursue the matter further by constructing an averager, but a functional instrument was not obtained until 1959. This instrument, similar in principle to Dawson's (1954) summator, but more limited in its capacity to handle data, was constructed at the University of Iowa by Harold Shipton. It was

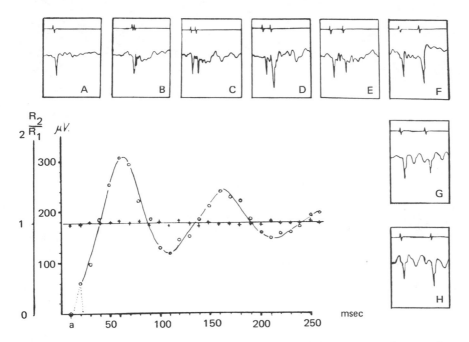

Fig. 3. Cortical excitability cycle obtained with paired flashes of light. The insets give examples of the responses from which measurements were made; interstimulus intervals range from 0 to 260 msec. Upper trace in insets indicates stimulus marker; lower trace, responses to flash. (Reprinted with permission from Gastaut *et al.*, 1951*a*.)

soon supplanted by a custom-made analog instrument with greater capabilities (Shipton, 1960) and, finally, by commercially produced digital averagers. Most recently, the analysis of our data has been facilitated by the use of general-purpose computers to process the output of digital averagers.

In our research to date, we have conducted more than 2000 experimental sessions with psychiatric patients of all kinds, in addition to several hundred sessions with nonpatient subjects. The hypothesis that cortical excitability cycles would differ in varying psychopathological states was supported by the results of our initial studies, and these results generated additional questions that are still being explored. Although we were mainly interested in searching for neurophysiological correlates of psychopathology, it was necessary to devote much of our investigative effort to technical and methodologic problems. These were concerned with instrumentation, data reduction and analysis, and the determination of variables, such as age and sex, which could bias results. Looking back, there seems to be little doubt that the methodologic studies have so far contributed at least as much valuable information as work directed to substantive questions.

Chapter 2

Techniques

New developments bring frequent changes in the details of evoked-response techniques. Consequently, this chapter attempts to deal with techniques at a fairly general level, to indicate the needs and problems and some ways in which they have been met. The author's methods are described in some detail, because they bear upon research findings presented later; such detailed treatment should not be taken to mean that the methods described are being advocated as necessarily the best for the purpose. The reader will also notice considerable variation in the level of complexity of the material; it ranges from very elementary topics, such as electrode application, to rather difficult problems of data reduction. The selection of material has been governed to some extent by the questions encountered by the writer when he has been consulted by workers seeking to embark upon evoked-response research. It is hoped that the elementary portions will be of use to beginning investigators.

AVERAGING: CONCEPTS AND LIMITATIONS

The basic idea underlying the use of summation[1] to extract the averaged evoked response from the EEG has already been considered in Chapter 1.

[1] Although the term, "averaged evoked response" is in common use, averaging instruments, more often than not, yield the sum of a group of observations, or the sum plus a constant. Consequently, "sum" and "summation" would be more accurate terms than "average" and "averaging." However, since the mean is the sum divided by the number of observations, there is no essential difference between mean and sum for most purposes. For this reason, unless otherwise indicated, "averaging" and "summation" will be used interchangeably here, as will "averager" and "summator" and "average" and "sum." Also, it should be made clear that "average," as used here, refers only to the arithmetic mean or sum of a series of measurements, and not to the median or mode, except when all three measures of central tendency may be identical.

Figure 4 gives a simple illustration of the princtple. The desired signal is represented by wave C; it is one-eighth the peak-to-peak amplitude of sine waves A and B, which are shown on the left as completely out of phase with one another. The middle figure shows what waves A and B would look like if wave C were to be superimposed on each. One would have great difficulty in being sure about the nature of C in records where A and B are also present; this situation parallels the usual EEG. The figure on the right shows what happens when the two wave forms in the middle of the figure are added together. A and B add to a horizontal line which forms the background against which the summed C waves may be clearly seen.

Considering the evoked response as the signal and the EEG as random noise, the improvement in the signal-to-noise ratio (S/N) to be expected from averaging is proportional to the square root of the number of observations added. This means that the advantage in discerning the signal with averaging is relatively greater with a small number of observations; e.g., the gain in accuracy with only 16 observations is already half that obtained with 64. Another important advantage of a relatively small number of observations, is that it minimizes the likelihood that the condition of the brain will change during the averaging sequence. However, there is reason to doubt that the square root law fully applies to the usual averaging situation, because the background EEG is rarely, truly random. This probably reduces the number of samples required for a given amount of S/N improvement, because Walter (1964*a*) has shown that, if all of the background activity consisted of transients without coincidence, the S/N gain would be identical with the number of observations rather than with the square root of that number. In practice, the S/N gain probably falls somewhere between the square root and the total number of observations. The investigator must decide, usually empirically, upon averaging that number of observations which will provide the most reasonable compromise between a reliable, noise-free measurement, minimal risk of recording from a subject in more than one state and minimal expenditure of time.

It should be borne in mind that both random and systematic changes may take place during the averaging sequence. Figure 5 gives an example

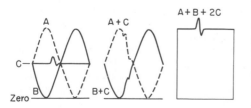

Fig. 4. Illustration of summation principle for extraction of small signals from larger "noise." The signal of interest is wave C. A and B are sine waves completely out of phase with one another. In the middle figure, C is superimposed on both A and B and is difficult to discern. When the waves in the middle figure are summed, they give the result at right; A and B add to a horizontal line and C is clearly apparent. (Reprinted with permission from Shagass and Schwartz, 1961*a*.)

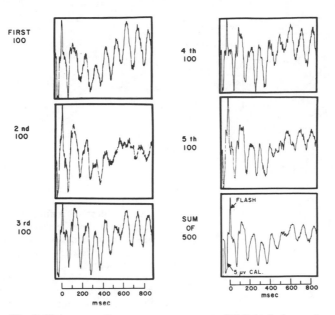

Fig. 5. Five consecutive average responses to 100 light flashes and average of all 500 (lower right). Note random variations in amplitude from one average of 100 to the next, which are not reflected in the sum of 500.

of relatively unsystematic variation in individual consecutive averages of responses to 100 light flashes in the same subject. The figure also shows the record obtained when all five averages are summed. Clearly, the average can conceal much variation that may occur during the course of the recording, some of which may be of interest to the investigator. If wave shape and timing remain constant, and only amplitude varies, the representativeness of the average may be satisfactory for many purposes. When shifts in time of occurrence (latency) of peaks take place, the average can be quite unrepresentative of the true course of events. An example of this is given in Figure 6 in which a consistent latency shift takes place from the first to the fourth average. No indication of this shift could be detected in the sum of the four averages (Brazier, 1964). The extreme case of latency variability involves a truly random temporal dispersion of signal peaks, so that the situation approaches one that would occur if waves *A* and *B* in Figure 4 were evoked responses; their sum would approach a horizontal line and no averaged response would be detected. Apparent reduction in amplitude of an averaged evoked response may represent variation in latency without actual amplitude change.

The desirability of obtaining an estimate of the variance of an average has led investigators with appropriate computer facilities to devise programs for obtaining such estimates. In scalp recordings, however, the evoked response may be only a fraction of the total electrical activity recorded;

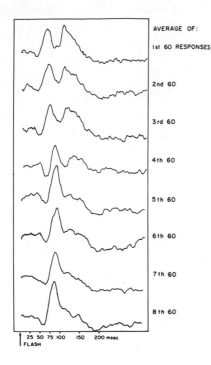

AVERAGE OF:

1st 60 RESPONSES

2nd 60

3rd 60

4th 60

5th 60

6th 60

7th 60

8th 60

25 50 75 100   150   200 msec
FLASH

Fig. 6. Effect of continued 1/sec flash on response in centre median. Successive averages of 60 responses demonstrate systematic variations in latency. The peaks in the first average occur earlier than those of the second, and so on. Although this effect is most pronounced in the first four averages of 60, there was no indication of these latency shifts in a single average of the first 240 responses. (Reprinted with permission from Brazier, 1964.)

thus, it would contribute only a small portion of the total variance, which would be hidden by the larger EEG variance. Furthermore, it is not possible to discriminate the variance contributed by the evoked response from that originating in the "spontaneous" activity. It has been suggested that an estimate of the contribution of EEG variance to the total variance can perhaps be obtained from EEG recordings taken in the absence of stimulation; this would seem to be a risky procedure, because sensory stimulation may alter the EEG in an inconstant manner, both from time to time in a given subject and from subject to subject.

GENERAL RECORDING CONDITIONS

The recording apparatus is usually outside of the subject room, which may be sound-deadened and electrically shielded, depending on requirements. One should be able to make the room light-tight and temperature should be maintained at about 75°F. Recordings are most often made with the subject sitting in a comfortable chair; for some purposes, he may be recumbent on a cot or bed. When the subject and instrument rooms are separated, an intercommunication system is desirable. It is convenient to apply electrodes in the subject room; when collodion is used for electrode attachment, a compressed-air outlet for drying it should be provided there.

Boredom is one of the principal problems encountered in recording; it predisposes to drowsiness or restlessness, both of which can affect results. The main cause of boredom is the necessarily long duration of the testing session in many experiments; provision for frequent rest breaks with opportunity to change posture is helpful. The recordist should also be aware of toilet needs, and it is good practice to give the subject an opportunity to void before electrodes are applied. Since incipient drowsiness is often denied by the subject, the EEG should be carefully monitored to ensure maintenance of a minimal level of wakefulness when the subject is passive with eyes closed. The EEG recording will also reflect excessive restlessness when it results in muscle potentials or the production of large transients due to movement. In our experience, individual averaging sequences should seldom last longer than 5 min; durations of 2 min or less are preferable. Although we have conducted numerous test sessions in which the subject was in the laboratory for three hours or more, experience amply supports the rule that, the briefer the experimental session, the better.

The degree of activity required from the subject varies considerably according to the type of experiment. In most of our work, we have requested the subject to be passive, with eyes shut, and with instructions to be as relaxed as possible while remaining awake. This type of recording situation maximizes what Sutton (1969) has termed "subject options." The subject can do whatever he wishes within the prescribed limits of muscular relaxation and wakefulness. Consequently, variations in mental activity are virtually uncontrolled and these may influence the evoked-response recordings. However, there have been no convincing demonstrations of relationships between a particular kind of self-selected mental activity, such as fantasy, and the electrophysiological measurements. Some workers, such as Callaway *et al.* (1965), have attempted to control the attentive state of the subject and, at the same time, to reduce artifact due to muscle activity by having the subject monitor his own EEG on a cathode-ray oscilloscope screen. It seems likely that this maneuver serves to reduce uncontrolled mental state variability. Even greater control may be achieved in experiments which demand specific voluntary responses from the subject, such as counting the stimuli, guessing which of several possible stimuli may occur, or responding to stimulation as rapidly as possible with a button-press. In addition to reducing subject "option," the required responses such as reaction time may also be measured and correlated with the physiological measurements.

Electrodes

Electrodes provide the essential interface between the subject and the recording apparatus and good electrode technique is vital for record quality. Since electrode application may occupy a substantial portion of the time spent in the laboratory by the subject, speedy methods are desirable, but speed must be balanced against reliability. The main consideration is to obtain a stable attachment with low resistance (5 kΩ or less) between two unpolarized leads. In addition to recording electrodes, a lead is required to ground the subject. There are several kinds of electrodes; these include metal discs, Bentonite paste, subcutaneously inserted needles, and pads made of absorbent material, such as felt. The method of attachment varies with the type of lead. We have preferred to use chlorided silver discs, made so that the central portion is raised to accomodate a conductive paste or solution, and with a hole in the center of the raised portion. Electrode paste is applied through this hole by means of a syringe with a blunt needle after the electrode has been stuck to the scalp by means of a small gauze square impregnated with collodion. The collodion attachment can remain stable for many hours and additional electrode paste can be applied without removing the lead. Electrode application may be facilitated by use of an applicator of the type manufactured by Specialized Laboratory Equipment, Surrey, England. This applicator has a tip which inserts part way into the hole in the electrode and holds it in place while the collodion is applied and dried. The applicator is connected to a compressed-air line; air is released through several small holes when a thumb-operated valve is opened.

Head locations for recording electrodes have not been standardized, although some workers have adopted the international 10–20 system developed for electroencephalography (Jasper, 1958). When a limited number of electrodes is used, the location is governed by the stimulus modality. Both scalp-to-scalp (bipolar) and scalp-to-reference (unipolar) derivations are employed. From a theoretical point of view, unipolar recordings are more desirable, but they are difficult to achieve. Favorite locations for reference electrodes are the ear lobe (singly or linked together), the mastoid, the bridge of the nose, the chin, and the zygoma. Unfortunately, there is uncertainty as to whether these locations afford electrically inactive references. Under many circumstances, high-voltage activity of extracerebral origin can be recorded from them.

Basic Instrumentation Needs

To record averaged evoked responses, amplified EEG samples must be entered into an averaging device. When enough samples have been accumulated so that the sum or average is likely to provide an accurate representation of the evoked response, it must be displayed. The essential equipment required then consists of: a stimulating device; an amplifying system with appropriate frequency and gain characteristics; an averaging device; a synchronizing or trigger signal to start the averager, either at the time of stimulation or at some fixed delay; and an instrument for displaying the average, preferably in some permanent form. Many instruments are available for each of these functions. The essential system may be greatly elaborated by the addition of extra timing devices, preset counters, calibration sources, instrumentation tape recorders for storage of the EEG, and equipment for storing the average in digital form and aiding quantification.

AVERAGING INSTRUMENTS

The devices that have been developed to achieve the purpose of automatic averaging or summation are of two general types: analog and digital. Digital devices have, for some time, been the most common type in use because they have been manufactured commercially, are more accurate, more convenient to use, and usually require less maintenance. However, analog types are still employed and some understanding of their capacities and limitations is desirable, since they have provided important data.

In evaluating an averager, the following characteristics are among the more important to take into consideration: reliability and accuracy of time base; resolution; range of analysis times; fidelity of retention of the store for display; permissible range of input voltages; intrinsic noise level; ease of display of average; and initial and operating cost. Time base reliability is of fundamental importance, because ideal averaging requires no time "jitter." Temporal resolution is determined by the number of data points (addresses) which the averager can store. Resolution requirements depend upon the rapidity and duration of the signal to be analyzed: high resolution is needed for signals with fast rise time, whereas for slow signals of equal duration the requirements are not as stringent. The desirable range of analysis times also depends upon the nature of the biological signals; very short analysis times are needed for recording averaged nerve impulses (Shagass and Schwartz, 1964a), whereas quite long analysis times are required for averaging slow potential phenomena (Walter, 1964b). If one wishes to average very fast and slow signals simultaneously in the same instrument, the number of data points must be large in order to resolve the rapid events. Linear accumulation of the signals to be averaged is most desirable; although nonlinearity can be handled in some cases by nonlinear calibration, this is troublesome and usually less accurate. Also, the number of samples that can be linearly summed is relatively restricted in many devices. In some averagers, the sum or average must be displayed quickly to avoid decay of the stored signal. In other instruments, such as the photographic averager, the process of displaying the average involves time consuming and costly operations. Ease and cost of display, as well as the form of the display, are of practical importance in determining the trouble and expense of operation. The permissible input voltages determine the amount of amplification required. The relation of the permissible input level to the level of noise intrinsic to the averaging instrument is important; the greater the ratio of permissible input level to irreducible noise level, the more noise-free will be the average.

Capacitor-Integrators

The summator described by Dawson (1954) employed a series of capacitors to sum the EEG at successive points in time. A rotating switch was used to sample the signal voltages at regular intervals after each stimulus. The display was on a cathode-ray tube, which was photographed (Figure 2). Dawson's instrument contained two banks of 62 capacitors which could be used to record either from two locations or earlier and later portions of the same events. The Dawson summator was used in a series of important studies; besides those of Dawson himself, these include investigations by Halliday and his coworkers (Halliday and Mason, 1964; Halliday and Wakefield, 1963) and Giblin (1964).

Barlow (1957) used a magnetic drum to determine the delay from stimulus of successive EEG samples to be summed by an integrator. This device required that the EEG be recorded on magnetic tape and that the tape be played through as often as the number of desired data points, corresponding to specified delay times. Operation was thus "off line" and

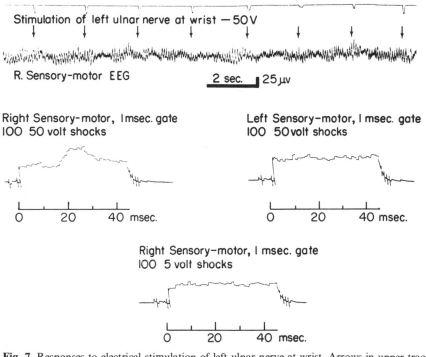

Fig. 7. Responses to electrical stimulation of left ulnar nerve at wrist. Arrows in upper trace indicate application of 50 v shocks; note absence of discernible effect on EEG. Summed recordings made with 44 integrator averager. With 50 v shocks, there is a clear response from right and absence of response from left sensorimotor area. With subliminal 5 v shocks, there is no response on right.

time consuming. However, the instrument provided for considerable flexibility in resolution, analysis period and number of observations.

The first averager used in our laboratory was similar to Dawson's summator (Shagass and Schwartz, 1961a). It consisted of 44 capacitor-integrators. The timing during storage of the signals was electronic, but an electromechanical switch was used for display. Figure 7 shows sample recordings obtained with this instrument. Resolution and analysis time could be increased by adding capacitors, but problems of size and maintenance impose a practical limit upon expansion of this type of averager. One contemporary averager (Princeton Waveform Eductor) operates on this principle. It contains no moving parts and switching is electronic. However, although relatively inexpensive, it has only 100 memory locations and can hold an average in memory for only a few minutes.

Photographic Devices

Photographic averagers use a photographic plate to store the EEG samples to be averaged. Photographic techniques were described by Calvet and Scherrer (1955) and by Kozhevnikov (1958). A two-channel instrument of this sort was constructed for our laboratory by Shipton (1960). The EEG was amplified and led to a modified two-beam cathode-ray oscilloscope. The oscilloscope beam was modulated so that fluctuations in brightness of the beam corresponded to the amplitude variations usually displayed on the Y axis. The vertical position of the beam was shifted slightly in a systematic order from the center to the periphery for successive sweeps. Pulse and waveform generators and a digital timer controlled the triggering of the averager and the timing of stimuli. The modulated beams were photographed on a Polaroid transparency, an example of which is shown in Figure 8. At the end of a series of stimuli, the transparency would contain light or dark bands (vertical in Figure 8) representing consistent fluctuations in voltage. The transparency was analyzed by means of an optical scanning device, in which the film was passed over a vertical slit through which a light was directed upon a photomultiplier. The output of the photomultiplier was written out with an XY plotter. Figure 8 shows such a writeout. It also shows some of the EEG from which the average evoked response was obtained, and demonstrates how difficult it is to discern the effects of stimulation in the ordinary EEG tracing. One other point in the figure to which attention may be drawn is the great regularity of the light and dark bands labeled T and P. In particular, T appears to be as regular in its occurrence as the thin white band representing the stimulus artifact. This means that the latency variability of the earliest component of response to ulnar nerve stimulation in this experiment was very small. It may also be seen that the later events of the evoked response display much greater latency variability than the regular early events.

The photographic averager offers rather good resolution for lengthy

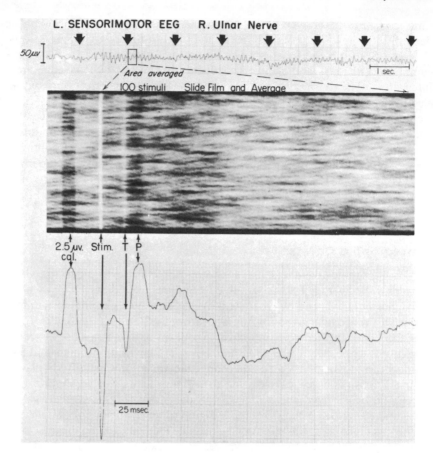

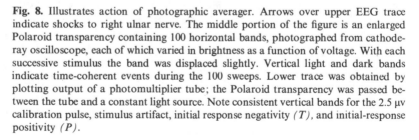

Fig. 8. Illustrates action of photographic averager. Arrows over upper EEG trace indicate shocks to right ulnar nerve. The middle portion of the figure is an enlarged Polaroid transparency containing 100 horizontal bands, photographed from cathode-ray oscilloscope, each of which varied in brightness as a function of voltage. With each successive stimulus the band was displaced slightly. Vertical light and dark bands indicate time-coherent events during the 100 sweeps. Lower trace was obtained by plotting output of a photomultiplier tube; the Polaroid transparency was passed between the tube and a constant light source. Note consistent vertical bands for the 2.5 μv calibration pulse, stimulus artifact, initial response negativity *(T)*, and initial-response positivity *(P)*.

analysis times. The number of signals that may be averaged is limited to about 200, which is adequate for most purposes. However, the method has several disadvantages. A major one is the nonlinearity of the response of the photographic emulsion to light. It is only with a great deal of difficulty that brightness and contrast can be regulated to the extent required to achieve linearity of response amplitude. Other problems arise from the lack of uniformity of enulsion between rolls of film, frequent presence of annoying defects, such as pinpoint holes, and the not inconsiderable cost of Polaroid transparencies. An additional handicap is that the optical scanning procedure must take place at a time other then recording. This introduces either delay or a need for an additional technician. Monitoring of the progress of the experiment may be difficult unless time is taken for optical analysis, although an experienced recordist can tell how things are going from inspection of the Polaroid transparency.

Storage Tube Averagers

Buller and Styles (1960) described an averager using the barrier grid storage tube. Storage tube averagers were used in important early studies conducted in the laboratories of Grey Walter (Walter, 1964a) and Henri Gastaut (Gastaut and Regis, 1965). Although Harold Shipton and John Emde built a storage tube averager for us, we used it for only a short time because a commercial digital computer became available.

Our brief experience with the storage tube instrument revealed several difficult problems. A few, such as the need to read out the store very quickly in order to avoid decay and some difficulty in maintaining a straight base line, were solved. However, a serious one was the fact that very precise adjustment was required in order to achieve the maximal function of the

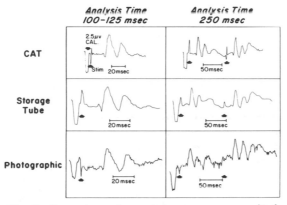

Fig. 9. Comparison of somatosensory responses simultaneously obtained with three types of averager. Upward deflection indicates relative positivity at active electrode. Two stimuli administered in tracings at right with 100 msec interstimulus interval. Responses recorded with short (100–125 msec) analysis times were similar with all instruments. With longer analysis time of 250 msec, storage-tube tracing shows less resolution; note smoothing of contours and loss of amplitude of initial downward component. Photographic averager record shows spikes resulting from minute holes in emulsion. (Reprinted with permission from Schwartz and Shagass, 1964.)

instrument. Although some workers have claimed that the resolution can be as high as 200 data points and that 200 responses can be averaged, the usual functional capacity of our instrument was well below this level. Figure 9 shows records obtained simultaneously from the same pair of leads in one subject with three different types of averagers: photographic, storage tube, and the Mnemotron CAT 400 Computer of Average Transients. Little difference can be discerned between the records when the analysis time is limited to about 100 msec. However, at an analysis time of 250 msec, the resolving capacity of the storage tube was clearly diminished while one of the difficulties with the photographic averager, namely, pinpoint holes in the emulsion, introduced small irregular "spikes" into the record.

Magnetic Tape Averagers

Some workers have used magnetic tape to store EEG samples for averaging. One instrument of this type was described by Rosner *et al.* (1960) and was used in important early studies from Rosner's laboratory. The essential principle was to add each new sample of EEG to the previously stored average by summing them together at the record head of a tape recorder. The sample size of the Rosner device was limited to about 125 observations.

Rémond's Phasotron

Since Rémond (1964) was particularly interested in spatial distribution of evoked and spontaneous activities, he designed an instrument that would do more than provide an average wave shape of evoked responses. The Phasotron combines analog and digital methods. Summation is analog, employing integrators. The average responses from each of several derivations, simultaneously tape recorded, can be obtained by summing voltage as a function of time. With the aid of a digital computer, single or average samples can also be processed by the Phasotron both as a function of time and as a function of topographic distribution along a line of recording electrodes. Figure 10 gives an example of the spatial–temporal map display form chosen by Rémond for this complex array of data. The digital computer classifies the potentials into categories according to voltage and polarity. The wavy lines are obtained by joining points in the same categories. The background for negative values is black; that for positive ones, white. The boundary between black and white is the point of reversal of phase.

Although more difficult to interpret than usual evoked-response curves, the data provided by the Phasotron clearly offer possibilities for a much more detailed evaluation of evoked-response activity than that afforded by the usual voltage-by-time presentation. It may be noted that even a map such as that of Figure 10 can represent only one topographic plane. A full

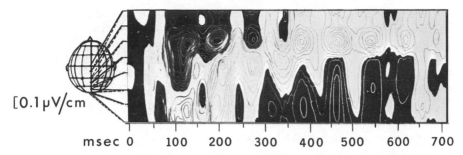

Fig. 10. Spatial–temporal map showing organization of responses to 100 light flashes at a period of 768 msec. Interelectrode distance is 2 cm. Potential differences transformed to gradients (μv/cm) as isogradient curves with a contour interval of 0.1 μv/cm. Abscissa, time; ordinate, distance. White areas, positive differences; black areas, negative differences. (Reprinted with permission from Rémond, 1968.)

topographic presentation, which would include both longitudinal and transverse planes, as well as progress in time, would require an additional dimension of display. In practice, Rémond displayed each topographic plane separately against time. Harris and Bickford (1967) have described a method for cross-sectional plotting of potential fields, that employs a fairly large digital computer (CDC 3200). The resulting graph is said to have three-dimensional properties which emphasize the phase relationships of the data.

Digital Computers

Digital computers are usually classified as either special or general purpose. In the special-purpose type, the desired analysis is wired into the control circuits; the program is thus fixed and difficult to change. In a general-purpose device, the programs are coded, and loaded into memory for use. They are easy to modify, and may be tailored to the user's specific requirements. If a limited function is to be performed, e.g., averaging alone, special-purpose computers tend to be less expensive and more efficient. Where averaging is only one of several desired computational procedures, general-purpose computers offer greater flexibility. Analog devices may have advantages over even special-purpose digital computers in well specified and limited applications but, as Clark (1961) has pointed out, digital methods become more desirable as the requirements of precision and stability increase. Furthermore, as the required analysis becomes more complex, it becomes difficult or impossible to execute by analog methods.

The first digital computer especially designed for evoked-response averaging was the Average Response Computer (ARC) designed by Clark and his coworkers at the Massachusetts Institute of Technology (Clark *et al.,* 1961). In addition to averaging, ARC was capable of compiling amplitude and time interval histograms. The operation of ARC in the averaging mode is as follows: The onset of the stimulus starts the action of the computer. After an initial delay, the computer performs an 8-bit analog-to-digital conversion at regular intervals, thus periodically sampling the EEG and generating a series of numerical values each accurate to one part in 256. Each value represents the electrical activity at a fixed time following the stimulus. It is added in the computer to a corresponding sum which is held in digital store. The sums are ultimately reconverted to analog form and displayed.

The basic principle of ARC was used in the construction of a number of subsequent commercially produced digital averaging devices. Among these are the Mnemotron Computer of Average Transients (CAT) and the Nuclear Data Enhancetron (ND800), both of which we have used in the work of our laboratory. Figures 5 and 9 show records obtained with Enhancetron and CAT computers, respectively. The resolution in time (abscissa) that is possible with digital computers is precisely known. The original model of the Mnemotron CAT contained 400 memory locations, divisible into two or four separate data channels or usable as one. A new model (CAT 1000) contains 1024 locations, similarly divisible. The Enhancetron ND800 contains 1024 locations, divisible into two channels. Data entry into consecutive channels of the CAT is not simultaneous, but sequential. There is thus a slight time discrepancy between channels, but this rarely causes difficulties. The amplitude resolution (ordinate) of these computers differs considerably because of different digitalizing principles employed in them. The Enhancetron has an 8-bit sampling accuracy at slow sweep rates; the signal is thus sampled with an accuracy of 1 part in $256 (2^8)$. However, at fast sweep rates the sampling error may be as high as 12%. The Mnemotron CAT averager uses the input signal voltage to modulate the frequency of an oscillator. The output of this oscillator is then counted to digitize the signal. Sampling accuracy is better when a higher oscillator carrier frequency is used; modulators are available with carrier frequencies from 100 to 1000 kHz. As with the Enhancetron, sampling accuracy varies with sweep rate and intrinsic noise level becomes a limiting factor in some applications.

Recent Developments

All of the averaging devices described above are more or less obsolete at this point due to changes in electronic technology. However, most of the evoked-response studies described in this book and found elsewhere in the literature used such instruments, and an understanding of their limitations may be necessary to correctly interpret the results obtained. Numerous new averaging devices have appeared on the market within the last few years, and these incorporate various improvements over the older instruments. Space does not allow a detailed description of the capabilities or limitations of each of these averagers, but some general trends can be noted.

Several instruments, including the Fabritek, Intertechnique, and Hewlett–Packard averagers, can perform a variety of analyses in addition to simple signal averaging; these include the computation of interval and amplitude distributions, cross- and autocorrelations, signal integration, and the measurement of evoked-response amplitudes. The most sophisticated current machine of this type may be the Time/Data 100 which can perform convolutions, Fourier transforms, and cross- or autospectral calculations in addition to the functions mentioned above.

Another group of recently introduced averagers uses integrated circuit memories in place of the ferrite core memories employed in most earlier devices. The devices that use integrated circuit memories may be less expensive than the Ferrite-core averagers. Some of these instruments provide a full range of data reduction functions and good temporal resolution; the Biomation 1000 is an example. Other, lower-priced instruments feature reduced memory sizes and more limited capabilities (Biomation 100). Another new approach is incorporated in the Technical Instruments Corporation CAT model 747. This instrument functions somewhat as a general-purpose computer in that its functioning is controlled by a program which is read into part of memory, the remainder of memory being available for data. However, the actual versatility of this instrument will, in practice, depend on the number of control programs developed by the manufacturer. It is not yet clear how many such programs will be developed, or what their cost and utility will be. With such an instrument, it is not practical for the individual user to write his own programs as he might with a truly general-purpose computer.

Small general-purpose computers are making their way into laboratories in which evoked responses are recorded. This development was viewed as an ideal to be striven for by Clark (1961). He compared the TX-O, a relatively small, but powerful, stored-program computer, with the ARC. He pointed out that the TX-O could be programmed in a matter of hours to act like the ARC. However, in addition, it could generate varied displays of data and could be programmed to carry out very lengthy and complex operations. Furthermore, the investigator could use a general-purpose computer "on line" to obtain calculated results while the experiment was in progress; he could then act on the basis of this information.

Computers incorporating the LINC instruction set have been popular in evoked-response laboratories, and the capabilities of the LINC computer appear well suited for EEG analysis. Various LINC computers have been manufactured, with the two currently available models being the Spear Micro-LINC and the more recently introduced PDP-12, manufactured by Digital Equipment Corporation.

For the investigator of human evoked responses, the small general-purpose computer clearly offers numerous possibilities for both "on" and "off" line analysis of data. If a store of tape recorded analog data is available, new mathematical models for reducing them may be applied long after the original data were taken, without need to repeat the experiments. However, such advantages can only be obtained from a small general-purpose computer if a major effort is devoted to program writing. The size of the investment in programming required to truly utilize a general-purpose computer can hardly be overestimated. Some computers are sold with "software" sufficient to allow them to do simple averaging and histograms. Nevertheless, the capability of such systems is no greater than that of many hard-wired averagers. General-purpose computers are usually more expensive to maintain than are special-purpose computers, and the evoked-response investigator can apparently profit from using such a machine only if he is prepared to invest a substantial effort in developing programs specific to his requirements.

ADDITIONAL INSTRUMENTS FOR RECORDING

Of the various instruments grouped under this heading, detailed attention will be given here only to those that have been specially developed for evoked-response work. The remainder will be considered only in terms of general function and requirements.

Preset Counters

In much evoked-response work, the number of observations is made standard for an experiment. The counter is set to the number of observations desired and circuits are arranged so that when the counter reaches this number, the stimulator will no longer be triggered and no data will enter the averager.

Timing Devices

These are required to initiate and sometimes terminate stimuli, to set the averager into action, and to control the speed of some display devices. Most averagers and stimulators have controls for duration of action and need only to be started by an external signal. Complex stimulus sequences require fairly elaborate timing devices. Commercially produced digital timers, which offer great precision and considerable flexibility, are available. With special-purpose digital computers, such as the Mnemotron CAT, it is possible to use the time base of the computer to obtain precise timing of events during the averager sweep. Random, rather than regular, presentation of stimuli is often desirable; this can be accomplished by means of a random-interval generator inserted into the timing circuit.

Calibration

Time calibration usually presents no problem, especially with digital equipment. Amplitude calibration requires special devices. Emde's (1964) low-level calibrator was designed for our laboratory and we have used it effectively with several types of averagers. It permits insertion of a calibration signal, selected to be in the amplitude range of the expected evoked response, in series with the recording electrodes. The calibration signal, examples of which are shown in Figure 9, is treated in the same way as the evoked response. Changes in the system from the recording electrodes on are reflected in the calibration. Gartside *et al.* (1966) have used a calibration device depending upon a photosensitive switch, which seems to be as effective as the Emde calibrator.

Total system calibration is particularly desirable for averaging procedures, since consistent small errors may sum. Some amplifiers require specially careful adjustment of dc balance and in-phase discrimination controls. Without such adjustment, positive and negative signals are not equally represented. When using such amplifiers, we have employed a routine checking procedure, using a device similar to the calibrator of Gartside *et al.* (1966). A battery voltage is attenuated to 5 µv. To ensure complete isolation of the test signal from ground, the battery is turned on by a photosensitive switch responding to a light source activated by a square wave pulse originating in a pulse generator. The battery voltage is entered at the input of the EEG, amplified and stored in the averaging computer. After the desired number of signals has been accumulated, the polarity of the input lead is reversed and the same number of signals is again entered into the computer without erasing the original store. If the system is properly balanced and adjusted, the positive and negative test signals sum to a horizontal line. If this does not occur, further adjustment is required. By generating the test signals in the time sequence of the stimuli to be used in the experiment, it is often possible at the same time to check the time settings for the experiment. This checking procedure permits one to be assured of the condition of the entire recording and timing system within a few minutes.

Amplifiers

The EEG must be amplified to the voltage level optimal for the averager. The number of stages of amplification will depend upon the averager requirements. To avoid exceeding the saturation level of the averager, gain controls which permit adjustment of the degree of amplification are desirable. Resistance–capacitance coupled (ac) amplifiers are in common use, although direct-coupled (dc) amplifiers are desirable on theoretical grounds. However, the latter may present difficult problems of base-line drift. The desirable time constant, which represents the amount of time taken for a level voltage deflection to be reduced by 63%, depends upon the characteristics of the evoked potential; for the usual evoked responses, it should be 0.2 sec or longer. For slow potentials, like the contingent negative variation (Walter *et al.,* 1964*a*) it should be at least 5 sec. The upper frequency cutoff should also be selected in accordance with evoked-response characteristics; to avoid distorting or attenuating rapid events, we have followed the practice of setting this control 3 kHz or higher. The frequency response of some amplifiers in clinical EEG instruments becomes nonlinear well below this range. In connection with upper frequency response of amplifiers, it should be recognized that the "notch" filter provided in some polygraphs, which is designed to reduce 60 Hz "hum," will, if used, also affect the biological signal. Difficulty is encountered in attempting to use amplifiers which employ mechanical choppers in their circuit, because the frequency range of amplification is limited usually to about 100 Hz; choppers also tend to introduce transients into the system. If a polygraph or clinical EEG instrument is used, its inkwriting system can be employed to monitor the EEG during evoked-response recording. However, since the frequency response of the pens will rarely be adequate for displaying rapid events in detail, concomitant display of the EEG on a cathode-ray oscilloscope is helpful.

Instrumentation Tape Recorders

Many research purposes can be achieved only if the EEG data are stored on analog tape. Analog tape storage can be used when there is no averager available for "on line" recording, or to augment the channel capacity of an available averager. For example, if one wishes to record evoked responses from ten head locations simultaneously, and the averager has only four data channels, recording from six additional lead derivations can be placed on tape with a 7-track recorder. The seventh track can be used to record trigger signals for activating the averagers when the data are reproduced. Within limits, the taped record can be reproduced faster than it was recorded to reduce playback time. However, faster reproduce speeds reduce the sampling time available to the averager for a given amount of data; since relative noise level increases when sampling time decreases, this sets practical limits for reproduce:record ratios. With the Mnemotron CAT, the use of high frequency modulator cards, e.g., 500 or 1000 kHz instead of 100 kHz, can permit higher reproduce:record ratios than would otherwise be possible. Perhaps the most important advantage of having data tape recorded is that they can then be analyzed in several different ways; also, new ideas for analysis may be applied to previously stored data without need to carry out another experiment.

The record speed employed should be selected in relation to the frequency range of the biological signal. For example, one recorder provides a very slow speed with a high-frequency cutoff at 100 Hz. This frequency response is adequate for EEG recording, and the slow record speed is ideal for making all-night sleep EEG records. However, the 100 Hz frequency response is insufficient for much evoked-response recording and would limit resolution of the signal well below the capacity of the usual averager.

Tape recorders can be subjected to remote controls which start and stop them automatically as part of a timing sequence. In addition to trigger pulses for starting the averager, various codes may be placed on the tape to identify the conditions under which the EEG was recorded so that when the data are retrieved, EEG samples may be grouped as desired in the averager. Several kinds of coding systems are possible; codes depending upon different voltage levels or pulses that vary in number are in common use. The recent introduction of integrated circuits has made it possible to construct coding equipment capable of generating complex identifying information in a precise manner at moderate cost.

In addition to analog tape recorders, digital recorders are available. These can be used to store the output of many averagers in digital form for entry into a compatible general-purpose computer which is programmed to perform measurements upon the data. Since data can be entered on digital tape in a matter of a few seconds, a digital recorder can also be used to store the "on-line" output of a digital averager quickly during the course of an experiment. This will substantially reduce the time the subject must spend in the laboratory, as plotting can consume several minutes between averaging sequences.

Display Instruments

Evoked responses are most commonly displayed in analog form as a graph of voltage fluctuation in time. The time base is usually linear, although some workers have used a logarithmic base. The evoked-response curve may be displayed on a cathode-ray oscilloscope and photographed, as was common in earlier work; although accurate, this involves the inconveniences and time delays of photography. Averaged responses may also be displayed on paper employing the ink writing unit of the polygraph or EEG recorder. However, the records obtained with these instruments are usually rather small in size and, unless the recorder provides for a rectilinear display, distortion is introduced by the curvilinear arc of the pen. Perhaps the most commonly used instrument for display is the XY plotter, which provides a fairly large curve on a sheet of graph paper without the problems of curvilinearity (Figure 9). Digital averages generally provide an adjustable time-base generator for the XY plotter. A plotter with a fixed time base driven by a motor (TY) may give equally satisfactory results. In general, the purchase cost of a plotter is correlated with the maximum possible speed of pen movement and sensitivity. Another cost factor is that of the recording paper; a difference of a few cents per sheet may be substantial when large numbers of graphs are made.

With some digital averagers, it is possible to purchase auxiliary instrumentation that will print out the numerical values held in each successive memory location or punch the digital values on paper tape. These forms of display may be much more convenient for quantification procedures than the analog graphic plot. The possibility of recording the digital store on magnetic tape has been mentioned previously.

Special Logic Circuits—Automatic Subtraction

Depending upon experimental needs, special logic circuits may be arranged not only to time stimulus sequences but also to deposit the EEG samples related to particular stimuli in different portions of the computer memory. In our experiments, we have made considerable use of circuits for automatic signal subtraction to assist in the measurement of recovery functions. In measuring recovery, one wishes to obtain separate measurements of the responses to the first and second of a pair of stimuli. With short interstimulus intervals, the second response (R_2) begins well before the first response (R_1) is completed. The record of the responses to the paired stimuli will therefore show the second response superimposed upon the first (Fig. 11). To obtain an estimate of the probable size of R_2 alone is a tedious task by hand measurement; it involves recording the averaged response to an unpaired stimulus (R_1) and measuring it at points identical in time to those at which the R_2 measurements are made for the paired response. The values for R_1 are then subtracted from the R_2 values obtained

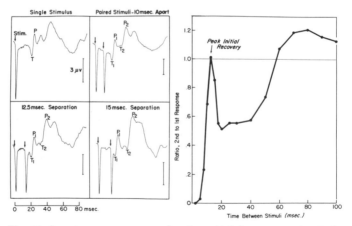

Fig. 11. Somatosensory recovery function. At left, responses to single and paired ulnar nerve shocks; upward deflection indicates relative positivity in active electrode. Amplitudes measured from T to P. Calibrations, 3 μv. Curve at right shows ratio of R_2/R_1 as a function of interstimulus interval. Plotted values are arithmetically smoothed. Deflections in single stimulus trace that occurred at locations T_2 and P_2 in paired response were subtracted by hand. Note initial peak of recovery before 20 msec, followed by period of R_2 suppression. (Reprinted with permission from Shagass and Schwartz, 1963a.)

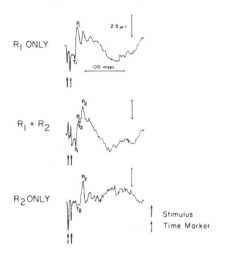

R₁ ONLY

R₁ + R₂

R₂ ONLY

Stimulus
Time Marker

Fig. 12. Illustrates function of automatic response subtractor. Single and double stimuli were alternated. R_2 obtained by subtracting R_1 from $R_1 + R_2$. (Reprinted with permission from Schwartz and Shagass, 1964, *Annals of the New York Academy of Sciences,* volume 112, pages 510–525. Copyright 1964, New York Academy of Sciences.)

from the paired response. All such subtraction involves the assumption that R_2 is simply added to R_1. This assumption appears to be generally correct, although the possibility that the second stimulus affects the later portion of the first evoked response cannot be entirely excluded. However, hand subtraction involves additional assumptions that are often not justified. One is that R_1 is similar from one averaging sequence to another. Another is that the R_2 peaks can be accurately visualized in the paired record. Since it is clear that the best R_1 estimate for subtraction is obtained from records taken as close as possible to the administration of stimulus pairs, we employed an automatic subtractor circuit which alternated the presentation of paired and unpaired stimuli within a sequence (Schwartz and Shagass, 1964).

The circuit was arranged so that the EEG associated with paired stimuli would be entered into one channel of the averager, while that related to unpaired stimuli would be entered into both channels. The polarity of the unpaired response, entered into the channel also receiving the paired response, was reversed, so that this channel contained $(R_1 + R_2) - R_1$ and thus yielded an estimate of R_2. The function of the subtractor is illustrated in Figure 12. It is noteworthy that measurements from T_2 to P_2 in the unsubtracted record would have yielded a value for R_2 much smaller than that obtained in the subtracted record. This is because R_2 latencies may differ from those of R_1 during the course of recovery, and such latency differences may not be detectable in unsubtracted records. The automatic subtractor circuit can be arranged to operate from data stored on a single channel of analog tape if an appropriate code is placed on the tape.

Artifact Suppressor Circuits

Since large transients due to movement artifact, eye blink, or high-voltage EMG (electromyograph) can unduly distort an average, circuits have been developed to eliminate those samples of EEG likely to contain them. In essence, these circuits detect signals above an arbitrarily established amplitude and operate to inhibit the stimulator or averager trigger for a predetermined time after these signals occur. Satterfield (1966) has described such a circuit for use with taped data. The artifact supressor used in our laboratory, designed by Donald Overton, has a nonlinear frequency response; it can respond to high- and low-frequency artifacts while ignoring EEG alpha frequency signals of equally large amplitude. Other devices do not have the differential frequency feature. Different methodological decisions have to be made when the artifact suppressor is used "on line" or with taped data. If used "on line," the duration of the averaging sequence may be lengthened because stimuli are blocked; however, the number of observations can remain uniform. With taped data, containing a uniform number of EEG samples in a set time, artifact suppression will cause variation in the number of EEG samples accepted for averaging. Despite these drawbacks, artifact suppression is very useful in studies of disturbed or uncooperative subjects from whom it is impossible to obtain quiet relaxation for an entire averaging sequence.

Automatic dc Base-Line Reset Circuit

With dc or long time constant ac amplifiers, used to record slow potential phenomena, a particularly troublesome problem is provided by base-line drift. This may reduce the dynamic range of the recorder or produce a saturation effect. Furthermore, when saturated by large transients, such amplifiers may remain unresponsive for a long time. These problems are accentuated when more than one amplifier is used. To cope with them in our laboratory, we use a circuit which automatically resets the base line to zero at specified times (Straumanis et al., 1969a). When averaging, the reset time is selected to fall between averager sweeps, so that the artifact associated with resetting is not visible in the average.

Dynamic Ground

Signals from the brain are much smaller than other signals which can be induced in the subject's body by interference sources such as 60 Hz mains and radio stations. EEG amplification is always accomplished with differential amplifiers, and the common mode rejection ratio specifies the degree to which these amplifiers will tend to reject unwanted interference signals. The amplitude of the interference signals induced on the subject's body can also be somewhat controlled by connecting one electrode to ground. However, due to electrode resistance, this does not reduce the common-mode artifact signals to zero, and if recordings are made outside of a screened room, or if the artifact signals "leak" into a screened room, 60 Hz "hum" may be observed in the resulting averages. We have found a device called "dynamic ground" very useful in controlling such artifacts. The dynamic ground is a negative feedback system which will drive the electrical potential of the subject's body very close to zero volts irrespective of any interference sources which may be present in the environment. With the dynamic ground there is no "ground" electrode, strictly speaking. A "sense" electrode is applied to the subject, and the voltage recorded at this electrode is compared to true ground. Any artifact voltage detected by the sense electrode is inverted in polarity and amplified. This provides a correction voltage which is connected to the subject's body through another electrode. Such circuits, which have been incorporated in some EEG amplifiers, have been described in the literature (Kopec, 1967), and in our experience are very useful. They greatly improve the quality of records taken outside of a screen room, and even if a screen room is available, they save time which would otherwise have to be spent looking for and eliminating artifact sources. They also reduce the stimulus artifact produced by electrical nerve stimulation.

STIMULUS PROBLEMS

For reliable results, it is necessary to use accurately specified stimuli that are the same in a given subject at different times and from one subject to another. These requirements are difficult and, in some ways, not always possible to meet. Even though the stimuli may be truly standardized in the physical sense, and effective techniques are employed to avoid pitfalls, such as changing pupil size with visual stimulation, great variation may be introduced by the need to repeat large numbers of identical stimuli in the averaging sequence. With simple stimuli, habituation may occur; with complex or subtle stimuli, meaning may change rapidly as a result of repetition. The latter limitation is unfortunate from the point of view of psychiatric interest in the relations between stimulus meaning and response.

Electrocutaneous Stimulation

Most of the work in our laboratory has been done with responses evoked by electrical stimulation of the skin over the ulnar or median nerve at the wrist. Figure 13 shows our arrangement of electrodes for stimulating the median nerve. The stimulating electrodes are the same kind as those used for recording; they are chlorided silver discs applied 3 cm apart with the cathode proximal. The ground electrode, usually a 1 inch square chlorided silver plate, is attached with tape to the stimulated arm 1 or 2 inches

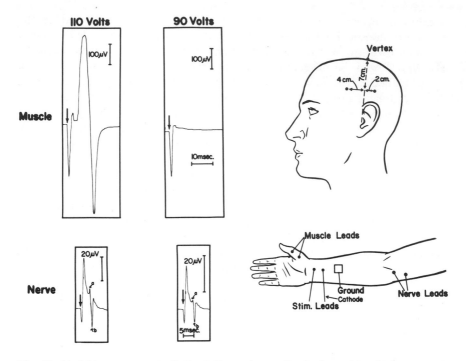

Fig. 13. At right, arrangement of stimulating and recording leads used to obtain averaged responses to stimulation of median nerve at wrist from cortex, median nerve at elbow, and muscles in thenar eminence. Tracings at left show large muscle response when stimulus was 110 v and absence of such response to 90 v shock. Summated nerve action potentials in lower left tracings were greater with stimulus intensity giving motor response. (Reprinted with permission from Shagass and Schwartz, 1964a.)

proximal to the cathode. This placement of the ground appears to have some value in reducing the stimulus escape artifact.

Our stimuli have been square wave pulses of brief duration (0.1 to 1.0 msec). Although we used a Grass S4 stimulator and stimulus isolation unit in our early work, we found that, at the voltage levels required, variations in subject impedance, coupled with the relatively high internal impedance of isolation units, result in inequalities between subjects of the actual voltage across stimulating electrodes. This could mean that individual differences in evoked responses might reflect nothing more than differences in actual stimulus intensity. To eliminate these inequalities, it is necessary to have either a constant-voltage or a constant-current stimulator. In order to decide which of these two kinds of devices would be preferable, a model of each was built and a comparative study was conducted (Schwartz *et al.,* 1964). In the same group of subjects, a full range of stimuli was applied twice with each stimulator in different order. The main criterion for comparing stimulators was the test–retest reliability of a single observation. When equating the intensity steps for the two stimulators, the reliability of a single intensity was 0.79 for the constant-current and 0.85 for the constant-voltage stimulator. At high stimulus intensity levels, the test–retest reliability was considerably higher, yielding a rank order correlation of 0.96. It thus appeared that constant-voltage and constant-current stimulators provided equally reliable data. We chose to use the constant-current stimulator because it is easier to manage when closely spaced paired stimuli are applied.

In practice, a constant-current generator is triggered and timed by a commercial stimulator, such as the Grass S4, through a stimulus isolation unit. Emde and Shipton (1970) have recently devised a digitally controlled constant-current stimulator which, under appropriate logic control, can provide stimuli of different intensities in one stimulus sequence.

Although the constant-current stimulator may control certain aspects of electrocutaneous stimulus variability, it cannot compensate for variation in the position of the electrodes and in the thickness and impedance of the skin and other structures over the nerve. In common with other workers, we have dealt with this aspect of the stimulus constancy problem by employing the sensory threshold as a physiological "zero" level. The justification for this procedure seems worthy of discussion.

Threshold is determined by decreasing stimulus strength until the subject no longer reports the stimulus and then increasing it from below threshold until he reports it. The mean of ascending and descending threshold values is taken as the sensory threshold. In our initial study of somatosensory evoked responses (Shagass and Schwartz, 1961a), we obtained data suggesting that the first measureable cerebral evoked response was found with a stimulus strength corresponding to sensory threshold. Confirmatory data were obtained in a subsequent study (Shagass and Schwartz, 1963a) of the relationships between stimulus intensity and response amplitude. In 25 subjects of a larger group, records were specially taken with stimulus intensities 1 to 3v below the descending threshold. In these subthreshold records, deflections between peaks at the appropriate latencies for the primary responses averaged 0.095 μv; they could be interpreted as random fluctuations about zero, due to noise in the system. In contrast, the mean amplitude of primary responses at sensory threshold in the same subject was 0.62 μv, significantly higher than zero and than the deflections obtained with subthreshold stimuli. The relationship between sensory threshold and the appearance of detectable somatosensory cerebral evoked responses has been further elaborated by Debecker et al. (1965). These workers obtained a careful psychophysical estimation of the sensory threshold along with evoked-response recording. They found that the primary evoked potential was first evident in response to a stimulus which was perceived by the subject between 10 and 50% of the time. They concluded that the statistical psychophysical threshold could be put into relationship with the evoked potential.

The data showing concordance between sensory threshold and the appearance of a measureable response at the scalp suggest that there is no

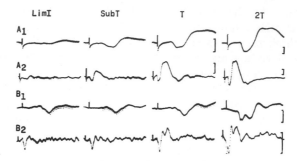

Fig. 14. Average evoked responses of exposed somatosensory cortex in relation to threshold stimuli at skin. Each tracing is the average of 500 responses at 1.8 per second. Total trace is 125 msec in A_1 and B_1, and 500 msec in A_2 and B_2; beginning of stimulus artifact has been made visible near start of each tracing. A and B, different subjects. Vertical column T: threshold stimuli, subject reporting no feeling of some of the 500 stimuli. Column $2T$: stimuli at twice threshold current; all stimuli felt distinctly. Column Sub T: subthreshold stimuli, none felt by subject; current about 15% below T in subject A, 25% below T in B. Column Lim I: subthreshold stimuli at "liminal intensity" about 25% below T in subject A, about 35 to 40% below T in B. Polarity, positive downward in all tracings. Calibrations, 50 μv in A_1 and A_2, 20 μv in B_1 and B_2. (Reprinted with permission from Libet *et al.*, 1967.)

cortical response to stimuli of subthreshold intensity. This conclusion appears to be incorrect in the light of findings reported by Libet *et al.* (1967). They studied somatosensory evoked responses in neurosurgical patients prior to operation. With carefully placed subdural electrodes, single shocks to the skin, which the subject could not perceive, evoked responses at the cortex. The early components of these responses to subthreshold stimuli were similar to those recorded when the subject experienced the stimulus, but the later components were missing at subthreshold levels. Libet *et al.* also found that, if subthreshold stimuli were delivered repetitively at 20 to 60 pulses per second for a minimum period of 0.05 to 0.1 sec, the stimulus was perceived. They showed that stimulation of the ventero-postero-lateral nucleus (VPL) of the thalamus and direct stimulation of the cortex could result in absence of sensation even though large primary responses were recorded from the somatosensory cortex. These observations led them to conclude that the later components of the cortical evoked potentials are better correlated with sensory awareness than the primary. Figure 14 shows sample records from somatosensory cortex obtained by Libet *et al.*, in which responses to definitely subthreshold stimuli are clearly visible. On the other hand, Libet *et al.* found that they too were *not* able to record an evoked potential with stimuli subthreshold for consciousness from scalp leads placed over the postcentral gyrus. Therefore, it appears that it is correct to assume that the first appearance of an evoked response at the scalp occurs with a stimulus intensity corresponding to the threshold for sensation, but that this does not mean that cortical responses are not elicited by subthreshold stimuli. It means only that such responses are not visible at the scalp or the dura (Domino *et al.*, 1964).

The foregoing statement is compatible with the results, although not with the conclusion reached by us, in an animal experiment (Schwartz and Shagass, 1961). In the cat, we demonstrated that the first recordable nerve action potential coincided with the appearance of the cortical evoked potential. Assuming, from our human data, that a cortical response required at least a threshold stimulus, we concluded that, in the conscious organism, subthreshold stimuli fail to excite either nerve or cortex and that this provides a physiological limit for "subliminal" perception. However, the results of Libet *et al.* suggest that both nerve and cortex may be excited by stimuli of subthreshold intensity.

Although the demonstration of subthreshold cortical responses raises some problems of interpretation, the uniform finding that evoked responses appear in scalp records at threshold intensity appears to justify the use of sensory threshold as a physiological "zero" reference point. Our procedure is to determine threshold and to scale all other stimulus intensities in relation to threshold employing the physical units, e.g., threshold plus 10 ma.

If response amplitude is plotted against stimulus strength, the resulting intensity–response curve resembles a growth function (Figure 15). Response

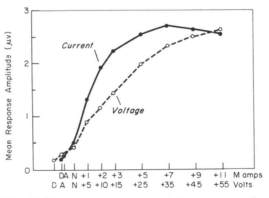

Fig. 15. Mean somatosensory evoked-response amplitude as a function of stimulus strength. Current refers to responses obtained with a constant-current stimulator; stimulus strength given in ma. Voltage refers to curves obtained with a constant-voltage stimulator for which intensity is given in volts. The same 12 subjects were used for both mean curves. *D* and *A*, descending and ascending sensory thresholds, respectively. *N* is nearest integer value in ma or multiple of 5 v above *A*. Succeeding intensity steps scaled in relation to *N*. (Reprinted with permission from Schwartz *et al.*, 1964.)

amplitude can vary greatly with small differences in stimulus strength in the rising portion of the curve, but will change relatively little over a wide range of intensities in the asymptotic portion. To maintain response stability during the course of an experiment, and to optimize amplitude, it is clearly advantageous to employ stimulus intensities that fall well along the asymptotic portion of the curve, e.g., 10 ma above threshold. We have followed this practice in our recovery function experiments. However, strong stimuli regularly elicit a motor response, e.g., thumb twitch to median nerve stimulation, and this raises questions about the effect of such motor phenomena on the cortical evoked response. In considering this issue, it may first be noted that it can be avoided by judicious selection of stimulus strength, since Dawson (1956) has shown that stronger currents are required to stimulate the motor than the sensory fibers of a mixed nerve. This is illustrated in Figure 13, in which averaged recordings of both the median nerve action potential and the electromyograph from the thenar eminence are shown. At 110v there was a large muscle response, which was absent at 90v; the nerve action potential at 90v was smaller than that at 110v, but still quite prominent.

When stimulus strength is sufficient to elicit a motor response, afferent activity arising from the muscle twitch can, in turn, produce a cortical response, so that the average response will reflect essentially two stimuli. However, it appears unlikely that the initial component of the somatosensory response is affected. Dawson (1956) found that the cortical potential is nearly maximal in amplitude at intensities lower than those required for EMG effects of the kind shown in Figure 13. He also found that the latency of EMG responses in the finger when stimulation was applied at the wrist was about 5 msec, so that proprioceptive impulses generated by the motor response would take more than 5 msec longer to reach the cortex than direct sensory responses to nerve stimulation. Thus, although the indirect proprioceptive volley arising from relatively intense electrocutaneous stimulation can certainly affect the later components of the response, it is not likely to affect the earliest ones.

Tactile Stimulation

The problems arising from electrical stimulation of a mixed nerve have led to attempts to devise stimuli that would be confined to the tactile modality. The most common procedure has been to use a blunt pin or stylus to make repetitive contact with the skin or with a fingernail. Shevrin and Rennick (1967) attached a stylus to the shaft of a solenoid mounted on a flexible stand. Walter (1964a) used a torque motor to drive the stylus. Giblin (1964) used a cork instead of a stylus; the cork was connected to a loudspeaker driven by a square wave generator. Giblin also used an air puff delivered by opening a solenoid on an air line. Figure 16 gives an example of the cortical response evoked by tactile stimulation. Comparison

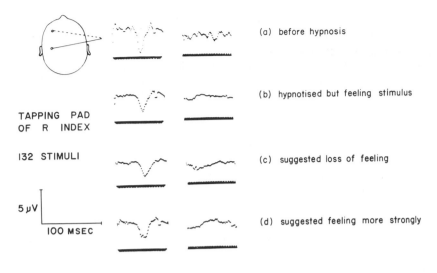

A.E. 31
28.10.61.

Fig. 16. Recordings of averaged evoked responses to 132 tap stimuli to the pad of the right index finger, given once per second. Record *d* was taken after it had been suggested that the stimulus would feel stronger, although stimulus strength actually remained unchanged. (Reprinted with permission from Halliday and Mason, 1964.)

with records in Figures 9, 11, and 12 indicates that this response tends to be simpler than the one evoked by electrical stimulation; this seems to be in accord with the idea that the different types of fibers in a compound nerve increase wave shape complexity. The main problem presented by tactile stimulation is one of controlling the strength of stimulus within a subject and standardizing it between subjects. Minor variations in the position of the stimulated surface can obviously affect stimulus strength. A practice, such as that used by Shevrin and Rennick, of covering the stimulated surface with tape may introduce additional intersubject variation. The intensity of air-puff stimulation would seem more readily controllable than that of a direct-contact stylus; although solenoid air valves may introduce "jitter," the averager can be triggered from a temperature sensor. It seems possible that further technical developments will lead to greater use of tactile stimulation.

Auditory Stimulation

Virtually no evoked-response experiments have been carried out under the precisely controlled acoustic conditions of a truly soundproof room with application of sound stimuli that are completely describable at the level of the eardrum. Auditory stimuli have usually been produced by activating a loudspeaker at a fixed distance from the subject's head or have been applied through earphones. Background noise has generally been dealt with by recording in a partially sound-deadened room and/or applying a consistent masking noise originating, for example, in a fan or air conditioner. The system described by Davis *et al.* (1966) is probably as sophisticated as any that have been employed in studies of the auditory evoked response. Their stimulating assembly can provide steady pure tones and/or white or filtered noise for masking purposes. It generates acoustic transients in the form of filtered clicks or "tone pips." A band pass filter is set for high- and low-frequency limits and is excited by a rectangular pulse from a Tektronix 161 pulse generator. The tone pips produced in this way reach their maximum voltage during the third cycle and subside almost as rapidly.

Many investigators have used clicks generated by exciting either a loudspeaker or earphones by means of a pulse generator or an electrical stimulator. The actual stimulus wave shape in these cases may differ considerably from the driving voltages. When earphones are used, the stimuli may be either monaural or binaural. When pure tones, produced by a sine-wave oscillator are employed as the stimulus source, it appears that the tone onset provides the effective stimulus for the auditory evoked response, since repeated cerebral responses are not seen when the steady tone lasts several seconds.

Some special problems associated with auditory evoked responses will be discussed more fully later. At this point, it may be noted that the initial events directly recordable from auditory cortex are usually not detected from scalp electrodes, and that most of the published work deals with later events with latencies of 50 to 300 msec. According to Davis *et al.* (1966), the repetition interval should exceed 6 sec if these later events are to be recorded when the cortex is fully responsive to each stimulus. The intensity and frequency of auditory stimuli has varied from one evoked-response study to another. There is general agreement, however, that very intense auditory stimuli are more likely to produce reflex muscle responses that may dominate the evoked-response record (Bickford *et al.,* 1964).

Visual Stimulation

More evoked-response work has been done in the visual modality than in any other. Most investigators have attempted to simplify the stimulus by using brief flashes, but even with these there are numerous problems of control. One instrumental factor is the nature of the flash tube used to generate the stimuli. In the wish to record the largest possible responses, relatively bright flashes have often been employed. A favorite instrument for generating such flashes has been a stroboscopic stimulator of the type used for studies of intermittent photic stimulation. The flash tube of such stimulators generally emits a click with each flash, so that a click-evoked response may be mixed with the response to the flash unless special precautions are taken. The problem of an accompanying auditory stimulus can be avoided by using neon lamps or glow modulator tubes. However, such lamps cannot generate a very intense flash.

With relatively bright flashes, evoked responses may be recorded with eyes shut; this may be mandatory to avoid serious discomfort. If the eyes are to be kept open, the intensity of the light may be reduced by increasing the distance between the lamp and the subject, or by using filters. Some investigators, e.g., Dustman and Beck (1963), have attempted to produce a uniform surround completely enveloping the subject's face by aiming the photic stimulator lamp from behind the subject onto a reflecting hemisphere of fairly large size, e.g., 70 cm diameter. Under these conditions a fixation point may be provided in the center of the hemisphere and recordings can be made with eyes open.

Some workers have employed stimulation with sine-wave-modulated light to study visual responses. This method was introduced by DeLange (1958). With it the light intensity varies sinusoidally around a certain value. When the modulation depth is increased, the difference between maximal and minimal light increases. This procedure permits one to increase the strength of stimulation without changing the average light level, thus keeping light adaptation constant. At a frequency of about 10 Hz, the threshold for perceiving a change in modulation may be as low as 0.5% (Van der Tweel and Lunel, 1965). When the frequency is increased, a greater change in modulation depth is required to obtain a subjective response. Spilker and Callaway (1969a) have recently employed sine-wave-modulated light to study personality correlates of visual evoked responses and effects of drugs.

The duration of the flash emitted by the usual stroboscopic type of photic stimulator is very brief, being of the order of microseconds. With such brief stimuli, the only response observed is that to the onset of the flash. However, Efron (1964) has shown that when flashes exceed 25 msec in duration, there is an "off" response so that the evoked response represents a combination of "on" and "off" components. With flashes lasting less than 25 msec, it is difficult to detect any "off" response. The color of light also affects response characteristics. Monnier and Rozier (1968) have studied the variations in evoked responses to white and colored light with low-intensity visual stimuli of relatively long duration. Various portions of the response showed changes in amplitude with different colors; variations occurred between conditions of light adaptation and dark adaptation, and these were also related to color. Clynes and Kohn (1967) have studied the evoked response effects produced by changes in intensity, color, and some form factors in visual stimuli. They used up to three 500 w projectors, sending the image onto a screen. The colors were produced by Wrattan filters and their projectors were equipped with high-speed shutters. They were able to demonstrate systematic variations of the responses in different areas with color, intensity, and some form characteristics of the responses. Effects of spatial patterning on the evoked response have been specifically examined

by several workers (Spehlmann, 1965; Eason and White, 1967). Some of
these experiments have involved the use of projectors. In others, such as
the one of Beatty and Uttal (1968), stimulus presentation was by means of a
cathode-ray screen controlled by a small general-purpose computer.
Studies which have aimed at discriminating responses of different meaning
have generally employed slide projectors for stimulus presentation (Lifshitz,
1966).

　　Two major factors that require control in visual studies are eye fixation
and pupillary diameter. An example of the importance of pupil size comes
from the work of Bergamini *et al.* (1965), who have provided evidence that
the habituation of the flash evoked response noted by many workers is

N. Lidia 42 y.

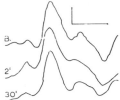

N. Vittoria 14 y.

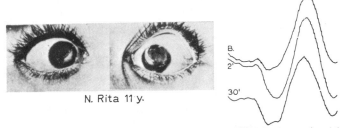

N. Rita 11 y.

Fig. 17. Averaged evoked responses to light flash in three subjects without
iris at the beginning and after 2 and 30 min of repetitive photic stimulation.
EEG samples are shown below each set of evoked responses. Calibrations
50 μv, 1 sec. Note absence of "habituation." (Reprinted with permission
from **Bergamini** *et al.,* 1965.)

primarily a pupillary, and not a cortical, phenomenon. Habituation is abolished by mydriatics and does not occur in subjects with congenitally absent pupil (Figure 17). The use of a fixed head position, such as provided by an optical stand with chin and head rest, together with a fixation target, serves to control fixation. Pupillary size variations may be controlled with an artificial pupil of a diameter smaller than would be achieved by maximal pupillary constriction. Although these procedures can be applied readily to cooperative subjects, they may not be suitable for psychiatric patients. Less cooperation is required when the size of the pupil is controlled pharmacologically, but installation of drugs may also involve difficulties in some populations. In addition to the stimulus, the surrounding illumination must be controlled. The simplest solution is to present the stimulus in a dark room, but some emotionally disturbed subjects may find a dark room distressing.

QUANTIFICATION AND DATA REDUCTION

Visual inspection of evoked-response curves obtained under different experimental conditions is rarely sufficient to test research hypotheses. Quantification of the data is required, particularly in psychiatric studies. Furthermore, since each response curve may reflect a large number of voltage measurements as a function of time, it is necessary to reduce these to relatively few representative values in order to achieve comprehensible results. The techniques employed for data reduction will depend upon the available instrumentation, the hypotheses to be tested, and the investigator's concept of the phenomena reflected in the evoked-response record.

Measurement of Components

If the evoked response could be assumed to reflect a single kind of neural event, quantification would be relatively easy; for example, the area under the curve would depict the activity. However, it is much more likely that the response curve is generated by several types of activity with differing temporal, and possibly spatial, characteristics. With no other evidence, this would be strongly suggested by the complex wave forms, with multiple peaks separated by variable time intervals. Consequently, investigators speak of different "components," usually defined by latency and polarity of visible maxima and minima in the curve (peaks). In considering quantification procedures, it seems important to inquire to what extent it is possible to measure the assumed components.

Figure 18 illustrates one of the commonly accepted models of how the evoked response is constituted. Figure 18A shows a series of positive and negative events, each with its own temporal course, but with overlapping times of occurrence. Thus, waves *a* and *c* of the graph are temporally

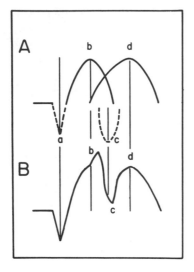

Fig. 18. The "component" model of a compound evoked response. *A*, four assumed components; *a* and *c*, negative; *b* and *d*, positive. *B*, algebraic summation of the components. Vertical lines show that peaks *b* and *c* are displaced to the right. Also, the negativity of *c* is not apparent in the synthesized response.

separate negative events, and *b* and *d* are positive events overlapping in time. Figure 18B depicts the curve obtained by summing events *a, b, c,* and *d*. It will be seen that the discrete negativity assumed to occur at *c* is not reflected as actual negativity in the curve. Furthermore, maximum activity associated with *b* and *c* is apparently displaced in time so that latency measurements of these peaks would not accurately depict these maxima. The evidence for the model shown in Figure 18A is indirect. Some of it comes from studies of drug effects which seem to demonstrate an unmasking of negativity not previously discernible in the undrugged state (Abrahamian *et al.*, 1963; Rosner *et al.*, 1963). Recovery function studies also support the model. With brief interstimulus intervals, components of the second response appear to be delayed and the delay is greater for later events; this has the effect of reducing overlapping of successive events so that they may be visualized. This is illustrated in Figure 19, which shows a series of tracings obtained from one subject in a study of somatosensory recovery functions. As the interstimulus interval was increased, successive negative peaks became prominent in the R_2 records. Peak 3, which is not visible in R_1, can be seen in R_2 at 2.5 and 5 msec delays; peak 5 is a distinct negative event with the same interstimulus intervals. Peak 7 becomes prominent as a negative wave in the R_2's obtained with intervals of 30, 50, and 80 msec.

Figure 18 demonstrates that only those individual events that do not overlap in time with others can be accurately depicted and measured in the evoked-response record. Since absence of overlapping cannot be assumed, with the possible exception of the very earliest events, it follows that amplitude and latency measurements based upon visible response peaks are probably quite inaccurate estimates of the characteristics of individual response components. Although aware of this problem, we have employed

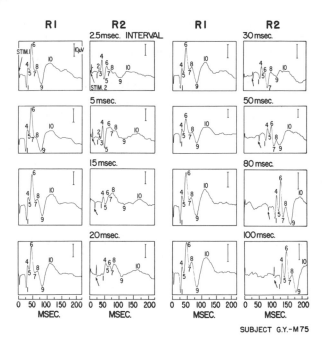

Fig. 19. Somatosensory evoked responses obtained in recovery function experiment. Normal subject, age 75. Relative positivity at active electrode gives upward deflection. R_1 is sum of 50 responses. R_2 is sum of 50 $(R_1 + R_2) - 50$ R_1. Numbers indicate consecutive peaks. Peaks 2 and 3 visible only in R_2 for 2.5 and 5 msec intervals. Note marked negativity of peak 5 at 2.5 and 5 msec intervals, of peak 7 at 20, 30, 50, and 80 msec intervals, and negativity of peak 9 at 50, 80, and 100 msec intervals.

such measurements in our studies, as have many other investigators, because they can be made without too much difficulty and they bear at least some relationship to the assumed underlying events. In attempting to determine whether these measurements were correlated with independent criteria, our rationale was that significant positive findings would indicate that the inaccuracies were not so great as to render the measures valueless. On the other hand, negative results would indicate either excessive inaccuracy, or absence of relationships.

Hand and Computer Measurement

It is probably correct to say that the majority of reports published so far have presented quantitative data based on hand measurements of latencies and amplitudes. For amplitude, the vertical distance between points on a graph is measured and converted to microvolts with reference to a calibration signal. Scaled measurements of horizontal distance from stimulus yield latency. There are two approaches to amplitude measurement. The vertical distance between designated maxima and minima provides so-called "peak-to-peak" measurements. The other approach is to measure amplitude deviations of peaks from an estimated zero base, or isoelectric, line. In our work, we have employed both approaches. The main problem with the peak-to-peak approach is that negative and positive acitivities may be combined into a single measurement when they should be considered separately. Inaccuracies in estimating the isoelectric line, due to noise or base-line drift, create serious difficulties for the second method. Neither method copes with the problem of overlapping components.

Computers have been programmed to yield, automatically, measurements comparable to some of those obtained by manual methods. Rodin *et al.* (1965), employed a "significant events" program providing amplitude and latency measurements. We have used an approach which divided the response into successive time segments, conforming roughly to the observable pattern of peaks, and determined latencies and amplitudes of positive and negative maxima within these segments (Shagass, 1967a). Computer quantification offers the obvious advantages of great savings in time and freedom from human error. However, it is extremely difficult to write a quantification program that provides for the full range of variations in evoked-response data and that can cope with the many possible artifacts. The problem is particularly great with respect to latency measurements, which depend upon accurate peak detection. When time ranges for differing peaks overlap, the technique of looking for maxima within a time segment is inadequate. Since the computer will deal with noise as data, noise peaks will be detected. This can be partially handled by "smoothing" the data first, but at the risk of distortion. In our work we have not yet been able to devise a satisfactory computer method for measuring latencies, except for that of the initial peak.

Our experience with amplitude measurements by computer has been more satisfactory. The output of our CAT 1000 computer is stored on digital tape which can be read by a general-purpose computer. Each record is preceded by a series of digits constituting a "tag word"; the digits identify the subject, the type of record, stimulus characteristics, lead placement, etc. Our amplitude measurement program employs the time segment approach; arbitrary and fixed portions of the response are quantified. The isoelectric line is estimated according to set rules. The deviation of a calibration signal of fixed amplitude from this line provides a scaling factor; this factor is used to convert the digital value of each data point to microvolts above or below the isoelectric line. The mean positive and negative deviations within each time segment can be computed. In addition, to obtain measurements independent of base-line variation, the average deviation from the mean for each time epoch is computed. The use of the mean value for a segment seems preferable to measuring peaks since the influence of noise is reduced. The computations are printed and also stored on magnetic tape for later statistical analysis. So far, we have found the average deviation value for the early somatosensory response events particularly useful. Correlations between automatically computed average deviation and hand measured peak-to-peak values yielded product-moment coefficients ranging from 0.80 to 0.96 (Shagass and Overton, 1970).

The use of general-purpose computers to automate data reduction according to hand measurement models, while valuable, hardly takes advantage of the potential for data analysis inherent in the computer. Several investigators have proposed, and presented data, using sophisticated statistical techniques for evoked-response analyses, which make greater use of the computer's capacity. Among these, mention may be made of the multivariate approach employing analysis of variance described by Donchin (1966) and the application of factor analysis by John et al. (1964) and Ruchkin et al. (1964). Methods such as these may ultimately provide descriptions of the evoked response in terms of numbers representing constants in equations relating amplitude to time and to source of activity. However, many problems must be solved before the computer can be used to full advantage. Those related to spatial distribution of response activity are among the most difficult.

Spatial Problems

Although scalp lead placements can be standardized with respect to external head markings, brain-to-scalp relations are not constant. This may be an important source of variability in measurements from a "standard" lead derivation. Some assurance that the measured record is being taken from optimal lead placements can be gained by recording simultaneously from several derivations. In the optimal record, the peaks to be measured will be of maximal amplitude. However, since the spatial distribution of peaks may not be uniform, no single derivation may be optimal for all peaks. This suggests that the records from several derivations should be measured and maximal values selected as representative, regardless of derivation. Although such a procedure may cope with some of the problems introduced by topographic variability, the fact that measurements come from different head areas in different subjects would still be a cause for concern. Another solution has been proposed by Rémond (1967).

Rémond has attempted to depict the activity recorded from eight bipolar electrode pairs chained in a line (Figure 10) in a single graph. When the eight individual evoked responses were algebraically averaged, the result was a virtually horizontal line (Figure 20a). However, when the successive amplitude values in each response curve were squared and the curves combined in a single graph of the root mean square, the result was entirely different. As Figure 20b shows, the root-mean-square curve revealed successive peaks of activity of variable amplitude. These peaks occurred at latencies in accord with those indicated in the individual curves and their height

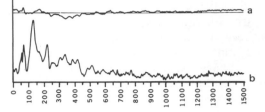

Fig. 20. Line a, sum of eight averaged responses to flash recorded in bipolar chain as in Fig. 10. Line b, root mean square obtained from the same recordings. (Reprinted with permission from Rémond, 1967.)

agreed well with impressions of relative magnitude. Furthermore, Rémond demonstrated that the root-mean-square curve shown in Figure 20b, which reflects recordings from nasion to inion, was very similar to one obtained from the same number of electrodes more closely spaced from vertex to inion, and also similar to a curve derived from a chain of leads transversely placed across the occipital area. These findings suggest that the root mean square obtained from multiple derivations may overcome some major difficulties introduced by spatial variability. However, although reduction of the root-mean-square curve to a relatively small number of amplitude values would be easy, computation of the curve itself is impossibly laborious without a general-purpose computer. It has not yet been used in studies of psychiatric interest.

Rémond's root-mean-square method essentially copes with the problem of spatial distribution by eliminating it. Satisfactory mathematical techniques that can be applied to reduce data in the space, time, and amplitude dimensions, simultaneously, have not so far been developed.

Fractionation by Frequency

Another approach to quantification is to assume that the complex evoked-response waveform can be fractionated into sinusoidal waves in different frequency bands. Since they are relatively uniform, the sinusoidal waves can be measured quite easily. They can be summed to provide an amplitude measure and their peaks can be easily located in time. Cazzullo *et al.* (1967) have employed this method. With band pass filters they divided the EEG into two frequency bands, namely 8–13 and 15–25 Hz. Figure 21 shows examples of averaged responses obtained with the unfiltered EEG and with EEG filtered in the two bands. It is seen that the filtered output consists of sinusoidal waves with relatively consistent intervals between peaks. As Cazzullo *et al.* point out, the problems associated with such filtering include a delay varying with the frequency band and oscillation of the filters when excited by a fast transient. Nevertheless, they were able to show that the digital data obtained by selection of their two frequency bands could, when added together, provide an approximate replication of the response obtained from the unfiltered EEG. The replication was, however, only approximate and there were some points of divergence. Better replication would undoubtedly have resulted from use of more frequency bands. The distortions introduced by instrumental filtering can possibly be avoided by using digital filters of the kind described by Ormsby (1961). With appropriate computer facilities, application of this type of quantification technique may provide useful and interesting results. However, it must be emphasized that the fractionation of evoked responses into sinusoidal waves, that can be recombined to yield the original wave form, provides no evidence that the potential originated from sinusoidal events. The fractionation technique is a way to describe and quantify the response, but does not necessarily reveal its actual constituents. The same may be said about Fourier analysis, which we used to quantify responses in a study of the effects of LSD (Shagass, 1967a).

Recovery Functions: A Special Problem of Quantification

The conventional method of displaying recovery functions in neurophysiology has been to plot the ratio of the second to the first response, i.e., the R_2/R_1 ratio. Figures 3 and 11 show curves plotted this way. One implication of using the ratio is that when $R_2 = R_1$, and the ratio is 1.0, full recovery of responsiveness has occurred. Another implication, which is of particular importance in studies of individual differences, is that variations in magnitude of R_1 from subject to subject will be equalized by the ratio measure, so that the differences in the degree of recovery can be directly assessed. Although we used the ratio measure in this way in our earlier studies (Shagass and Schwartz, 1961b; 1961c; 1962a), further examination of recovery function data indicated that the ratio is probably not an accurate measure of individual differences in recovery.

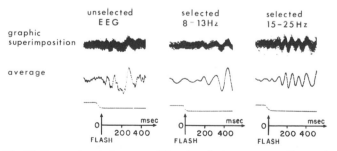

Fig. 21. Comparison between nonfiltered evoked-response readings and those obtained by filtering in selected frequency bands. Filtered responses show deflections with a more constant latency. (Reprinted with permission from Cazzullo *et al.*, 1967.)

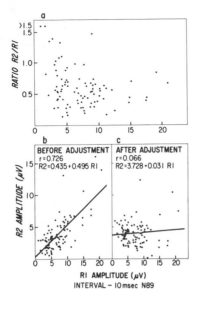

Fig. 22. Scattergrams illustrating problems of depicting recovery functions and a solution. Data from 89 nonpatient subjects. Somatosensory responses with a 10 msec interstimulus interval. a, conventional R_2/R_1 ratio as a function of R_1 amplitude; note high ratios with lower amplitudes. b, R_2 amplitude as a function of R_1 amplitude; regression line based on "within-groups" correlation, the groups being divided into four ranges by sex. Note high correlation and positive intercept. c, relation of R_2 amplitude to R_1 after adjustment for covariance shown in b. Note that the two values are now independent, permitting test of "between-groups" differences in R_2, independent of R_1. (Reprinted with permission from Shagass, 1968b.)

The nature of the problem is illustrated in Figure 22a; this is a scattergram relating R_2/R_1 ratios to the amplitude of R_1 for somatosensory recovery at a 10 msec interstimulus interval in 89 nonpatient subjects. It will be seen that, instead of being evenly scattered throughout the plot, as would be expected if the ratio actually compensated for variations in R_1, too many high ratios occur with low R_1 amplitudes. Figure 22b shows the scattergram obtained by plotting R_2 against R_1 amplitude for the same data. The regression line drawn through the data is the best-fitting one obtained by the method of least squares. Although the data fit the linear regression line quite well, it will be noted that the line does not pass through the origin, but that there is an intercept. Now, valid use of the R_2/R_1 ratio for examining individual differences demands that it meet two criteria: (1) that the regression of R_2 on R_1 be linear; and (2) that the regression line pass through the origin. Figure 22b shows that the second criterion is not met and it can be stated that, in dozens of distributions like those in Fig. 22b, we have nearly always found a positive intercept. This means that the variations in R_1 amplitude from subject to subject will introduce variations into the ratio. The problem may be depicted as follows:

$$R_2 = a + bR_1$$
$$R_2/R_1 = a/R_1 + b$$

Since a (the intercept) is a definite quantity, variations in R_1 will produce different values of a/R_1 in different subjects and affect R_2/R_1. In accord with Figure 22a, ratios would tend to be higher when R_1 is small and lower when R_1 is large.

The procedure that we have employed to deal with the problem is based upon analysis of covariance and was suggested to us by Dr. Dee Norton of the University of Iowa. Its main application is to compare different groups for differences in recovery. The method is extremely laborious without a large computer. There are three stages in the analysis. The first stage is to compute the *within-group* regression of R_2 upon R_1 for each interstimulus interval and for each group of subjects. In the second stage, the regression equations are used to adjust the R_2 values for their covariance with R_1. Figure 22c shows how the adjusted R_2 values vary randomly with respect to R_1 in contrast to the strong correlation present before adjustment, as shown in Figure 22b. The final stage of the analysis is to use the adjusted R_2 values to compare the significance of group differences by analysis of variance. It may be noted that, since adjusted values are used in such an analysis of variance, the number of degrees of freedom (df) must be reduced by one for each regression co-efficient.

Characteristics of Event-Related Potentials

Averaging methods can be used to record potentials other then those evoked by sensory stimuli. This has led Vaughan (1969) to propose the term "event-related potentials" (ERP's) to designate bioelectric signals that exhibit stable temporal relationships to a definable occurrence. Vaughan distinguishes five classes of ERP: (1) sensory evoked potentials; (2) motor potentials; (3) long latency potentials related to complex psychological variables; (4) steady potential shifts; and (5) potentials of extracranial origin. More than one class of potential may be present in any given recording and the distinction between classes may pose difficult problems. Studies of psychiatric interest have been conducted mainly with potentials in Vaughan's classes 1, 3, and 4; those in class 5 constitute biological artifacts that may contaminate results.

This chapter describes the various kinds of ERP, with particular attention to sensory evoked potentials. It should be emphasized that only relative descriptive statements can be made. The wave shapes, amplitudes, and latencies of ERP's vary greatly between individuals and are highly dependent upon location of electrodes, stimulus parameters, the experimental situation, and the physiological and psychological state of the subject. Thus, although it is common usage to speak of "the visual evoked response" or "the somatosensory evoked response," it should be explicitly recognized that the response evoked by stimulation in a particular sensory modality is by no means uniform.

SOMATOSENSORY RESPONSES

The cortical response evoked by electrical stimulation of a peripheral nerve is considered by Bergamini and Bergamasco (1967) to have shape and latency characteristics more constant than any other type of evoked potential. Although details vary, most published descriptions indicate that the response contains a series of wave peaks beginning about 15 msec after the stimulus and continuing for at least 300 msec. In our work, when we have needed to confine ourselves to only one lead derivation, we have employed the bipolar placement shown in Figure 13; it is similar to the one used by Dawson (1947). The posterior lead is assumed to be over the cranial projection of the somatosensory area on the hemisphere contralateral to the stimulated wrist (Krönlein, 1898). This appears to be an optimal placement when stimuli are applied to either ulnar or median nerves or to the fingers (Goff *et al.*, 1962). The optimal site is more lateral when the face is stimulated, and closer to the midline when the lower limb is stimulated. According to Halliday and Wakefield (1963), the impulses responsible for the cortical response travel by the dorsal column pathways. In neurological patients, they found that the response was markedly reduced when touch and position sensibility were affected, but not when pain or temperature sense was lost.

Description and Spatial Distribution

The distribution of the somatosensory evoked responses over the scalp was carefully studied by Goff *et al.* (1962), using 21 scalp electrode placements and a reference electrode placed on the bridge of the nose. Figure 23 shows recordings obtained with stimulation of the left median nerve. These authors designated successive components in the response by number; this type of designation differs from our practice of numbering successive peaks, in that a component may contain both positive and negative peaks. Figure 23 shows the early components clearly in the contralateral hemisphere, particularly at electrode $c4p$; this is 4 cm posterior to the location of our posterior lead in Figure 13, which is designated as cPR in Figure 23. The earliest components of the response (1, 2, and 3 in the scheme of Goff *et al.*) were very small or absent on the side ipsilateral to the stimulus. By contrast, the later components, 4 and 5, appeared to be bilaterally represented. These findings have been generally confirmed in a recent topographic study from the same laboratory (Goff *et al.*, 1969). The latencies of the Goff components were as follows: component 1, a positive–negative–positive sequence with peaks at about 16, 20, and 25 msec, respectively; component 2, positive, peaking at about 31 msec; component 3, positivity peaking at about 48 msec; component 4, negativity at about 65 msec and positivity at about 85 msec; component 5, negativity at about 135 msec and positivity at about 200 msec.

Goff *et al.* (1962) suggested that component 1 probably represents potentials in presynaptic thalamocortical fibers and that component 2 probably represents postsynaptic cortical potentials. They proposed that component 3 may be similar to evoked responses of "association" type

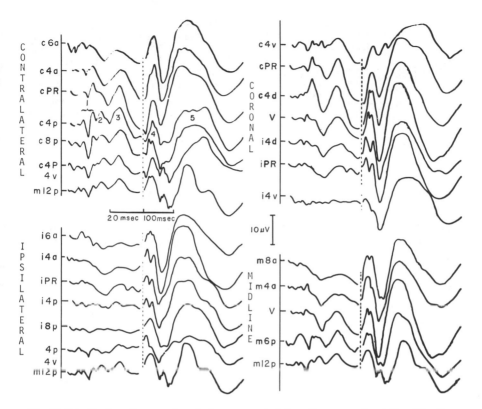

Fig. 23. Distribution of averaged evoked responses to left median nerve stimulation. The time base for the right half of the tracings is five times as long as that for the left half. Note that early components are visible mainly on the side contralateral to the stimulated wrist. (Reprinted with permission from Goff *et al.*, 1962.)

recorded in animals. Components 4 and 5 were considered possibly related to "*K*-complex" or "*V*-potential." Interpretation of the later components as reflecting extralemniscal activity, perhaps mediated by reticular formation, appeared to be supported by their bilateral occurrence in the recordings of Goff *et al.* (1962, 1969). The bilaterality would suggest a common subcortical source affecting the two hemispheres simultaneously. However, Domino *et al.* (1965) have shown, in man, that lesions in nucleus venteropostero-lateralis (VPL) reduced all somatosensory components during the first 125 msec after stimulation. Furthermore, Liberson (1966) found that, in patients suffering aphasia after cerebrovascular accidents, there was no trace of ipsilateral response to median nerve stimulation if the contralateral response was not observed. The findings of Domino *et al.* and Liberson support the view that the later, as well as earlier, somatosensory response events are mediated by specific sensory pathways. They suggest

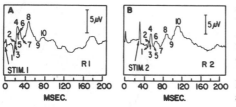

Fig. 24. Somatosensory responses to median nerve stimulation during recovery function determination. Interstimulus interval, 30 msec. Relative positivity at active electrode gives upward deflection. R_2 tracing obtained by subtracting 50 R_1 from 50 $(R_1 + R_2)$. Numbering scheme same as in Fig. 19. Note that points 5 and 6 are not apparent in R_1, but can be distinguished clearly in R_2.

that the ipsilateral later components may be transmitted from the contralateral cortex, and that reticular mediation is not directly reflected in the response. The data of Vaughan *et al.* (1968) concerning spatial distribution of somatosensory responses also suggest that the component corresponding to wave 5, of Goff *et al.*, is generated solely in the contralateral primary somatosensory projection area.

Figure 24 shows the numbering scheme that we have used to designate the first ten somatosensory peaks. Peak 1 is universally recognized as the initial negativity, interpreted by Goff *et al.* as a presynaptic volley. Peaks 2 and 3 would be included in the complex of component 1 by Goff *et al.* (1962). They are often not visible, being seen as distinct peaks mainly in young adults. Peak 4 corresponds to component 2 of Goff *et al.* Peaks 5 and 6 are frequently not apparent in responses to single stimuli, particularly in younger individuals; however, as Figure 24 indicates, they may be brought out in the response to the second of a pair of stimuli. Table I gives the mean amplitudes and latencies obtained for these peaks, together with standard deviations, in a group of 86 nonpatient subjects ranging in age from 15 to 80 years. The amplitudes were measured as deviations from an estimated isoelectric line. The latency values are somewhat greater than those reported

Table 1. Somatosensory Response Measurements in 86 Nonpatient Subjects

Peak[a]	Amplitude (μv)		Latency (msec)	
	Mean	SD	Mean	SD
1	− 2.27	1.54	20.2	1.74
4	4.70	2.92	29.1	2.62
5	2.77	1.90	33.4	3.49
6	5.02	3.60	38.7	4.33
7	1.54	2.40	47.8	5.13
8	3.33	2.26	58.1	6.44
9	− .01	1.32	73.9	8.85
10	2.75	2.22	94.1	12.61

[a] As in Fig. 24.

by us in an earlier study in which the average subject was 13 years younger (Shagass and Schwartz, 1964a).

Our own data on distribution are in essential agreement with those of other workers. Figure 25 demonstrates the predominantly contralateral representation of the response to median nerve stimulation. The figure also contains a comparison between recordings made with scalp-to-scalp lead derivations and with a monopolar arrangement using the two ears linked together as a reference. Although the monopolar and bipolar records were generally similar, there were some differences. The amplitudes of initial peaks were greater in the bipolar records. Also, close inspection of the monopolar recordings reveals a small positive event preceding peak 1, which is not seen in the bipolar traces. This event has been noted by a number of workers and is sometimes designated peak 0. However, it is not sufficiently constant or large to measure with regularity. Goff *et al.* (1962) noted it as the initial positive phase of their triphasic component 1. An event occurring even

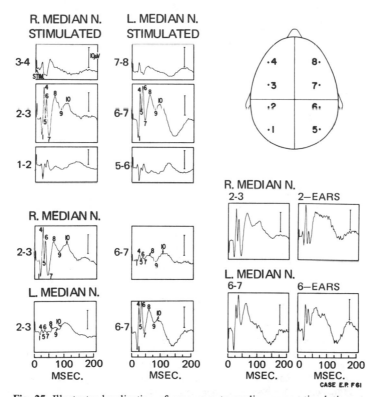

Fig. 25. Illustrates localization of responses to median nerve stimulation at electrodes 2 and 6, placed over post-Rolandic areas. Responses greatest on the side contralateral to stimulation. Tracings in lower right compare scalp-to-scalp recordings (2–3 and 6–7) with records made using the linked ears as reference.

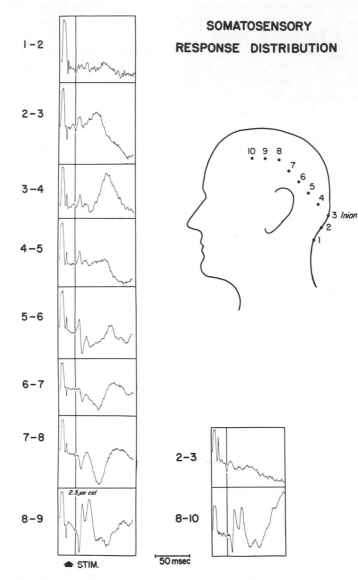

Fig. 26. Somatosensory response distribution. Electrodes 3 cm apart. Vertical line drawn through early wave with peak latency of 13 msec, which was maximally recorded between leads 2 and 3. This extracerebral wave may originate in the cerebellum or spinal cord. Leads 8 to 10 are customary bipolar placement when only one pair of electrodes is used. (Reprinted with permission from Schwartz and Shagass, 1964a.)

earlier than peak 0 and with a different localization, is shown in Figure 26, which was obtained by recordings from closely spaced leads. This wave, originally described by Liberson and Kim (1963), reverses in phase at the inion and has a peak latency of about 13 msec. It is possibly a cerebellar or spinal cord response. Cracco and Bickford (1968) were unable to affect it with muscle tension maneuvers, indicating that it is probably not myogenic.

Reliability

Somatosensory responses recorded from the same subject at different times under similar conditions tend to vary within restricted limits and to retain the same configuration. In an early study, we computed the correlations between the successive data points of two averaged responses taken 1 to 3 hr apart in time for each of 10 subjects (Shagass and Schwartz, 1961a). The mean coefficient was 0.87. In contrast, the correlations between responses of different subjects approached zero. These data lead to the conclusion that somatosensory responses under a given set of conditions are reasonably consistent and characteristic of an individual.

Effect of Stimulus Intensity

The latency of the somatosensory response depends upon the length of the conduction pathway and the strength of stimulus when state of alertness is constant. As would be expected, latency is greater when regions more distant from the brain are stimulated, e.g., the response to stimulation of the lateral popliteal nerve has a latency about 10 msec greater than that obtained by stimulating the wrist. Latencies are also greater in taller persons. Latencies tend to be longer with stimuli near threshold strength than with intense stimuli. The variation of amplitude with stimulus intensity is not proportional to the change in the stimulus strength. The intensity–response relations for the initial component (1–4 in our scheme) are shown for a group of subjects in Figure 15; Figure 27 illustrates the variations in amplitude of several evoked-response components with changes in stimulus strength for two individual subjects. It will be seen that the shapes of the intensity–response curves for different peaks were, in general, quite similar.

Recovery Functions

The normal recovery function of the cortical response to relatively intense electrical stimuli of peripheral nerve has been illustrated in Figures 11 and 19. Figure 28 shows a series of 11 individual recovery curves, which were measured in unusually great detail, since 20 interstimulus intervals were employed for the first 20 msec (Shagass and Schwartz, 1964a). Although amplitude recovery of the initial component varied considerably between subjects, the curves do reveal a general pattern of three phases: (1) some full recovery before 20 msec, with one or more R_2/R_1 ratios equaling or exceeding 1.0; (2) a variable duration period of diminished responsiveness, usually observable with interstimulus intervals between 20 and 30

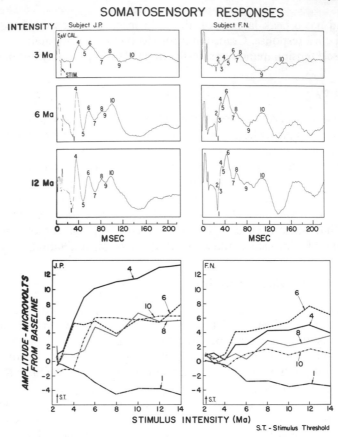

Fig. 27. Variation in amplitude as a function of intensity for different somatosensory response peaks in two subjects. Curves for peak 1 drawn to reflect negativity. Note the relatively uniform shape of curves for different peaks.

msec; (3) a succeeding rise of the curves toward a ratio of 1.0. The reliability of the peak recovery ratio of the initial component in 20 subjects tested twice several days apart was 0.78 (Shagass and Schwartz, 1962a).

The time course of latency recovery differs from that of amplitude recovery. Figure 29 shows mean latency-recovery curves for 6 consecutive peaks; the subjects were the same as those in Figure 28. (Peak 2 in Figure 29 corresponds to peak 4 in Figure 24, peak 3 to peak 5, etc.) The curves of Figure 29 show the mean latencies of R_2 to be prolonged at the shortest interstimulus intervals; as the intervals lengthen, the R_2 latencies become progressively more like those of R_1 and tend to reach the R_1 latencies with interstimulus intervals between 20 and 40 msec. Whereas amplitude recovery tends to be multiphasic (Figure 28), latency recovery tends to be monotonic.

The study yielding the data of Figures 28 and 29 was conducted primarily to determine the extent to which cortical response-recovery functions are dependent on those of peripheral nerve (Shagass and Schwartz, 1964*a*). Nerve and cortical recovery functions were measured simultaneously. Figure 30 shows the mean recovery function of human peripheral nerve; this seems similar to the curves obtained by Gasser and Grundfest (1936) for cat saphenous nerve. The nerve recovery ratios first equaled or exceeded 1.0 at intervals ranging from 3 to 8 msec (mean 4.5 msec). They then increased further to peak ratios averaging 1.30 at a mean time of 9.4 msec; the mean peak ratio in Figure 30 is less than 1.30 because peaks were reached

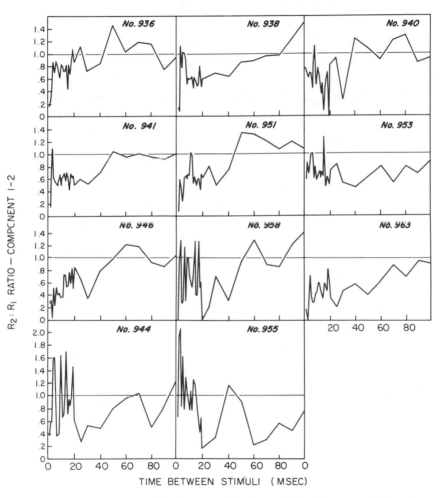

Fig. 28. Somatosensory recovery-function curves for eleven subjects. There were 20 interstimulus intervals between 1 and 20 msec. Amplitude measurements from peak 1 to the first positive peak, which corresponds to peak 4 in Fig. 27. Although the individual curves are irregular, most have one or more peaks with a ratio of 1.0 or more before 20 msec, a period of suppression between 25 and 40 msec, and the second peak of recovery after 40 msec. (Reprinted with permission from Shagass and Schwartz, 1964*a*.)

at different times in different subjects. After the peak, the ratios declined, dropping below 1.0 at a mean time of 14 msec. The period of diminished responsiveness persisted until about 100 msec. Correlations between the amount of recovery in the nerve and cortical responses were quite low;

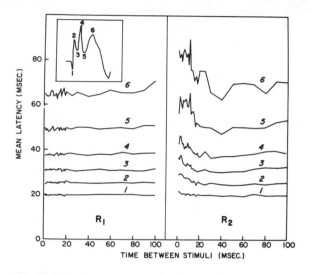

Fig. 29. Mean latency recovery for eleven subjects. Measurements are for each of 6 sequential points as in inset diagram. The numbering of peaks differs from that in Fig. 27 in that peak 2 here is peak 4 in Fig. 27, peak 3 is peak 5, etc. Note relative consistency of R_1 latencies and progressive decrease of R_2 latencies over first 20 msec. (Reprinted with permission from Shagass and Schwartz, 1964a.)

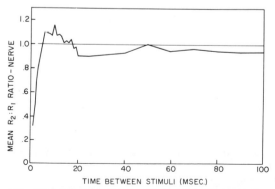

Fig. 30. Median nerve amplitude recovery curve for eleven subjects. (Reprinted with permission from Shagass and Schwartz, 1964a.)

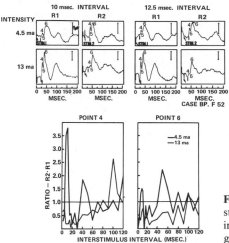

Fig. 31. Somatosensory recovery at two stimulus intensities in one subject. Conditioning and test stimuli of equal intensity. Note greater recovery with weaker stimulus.

they indicated that variations in nerve recovery did not account for more than 3% of the variance in cortical recovery. The major influences on cortical recovery appear to be of central origin.

Recovery functions vary in relation to stimulus intensity. This is illustrated in Figure 31, which shows sample tracings and recovery curves obtained with relatively weak and strong stimuli in the same subject. It will be noted that, in this subject, a psychiatric patient, there was suppression of R_2 when stimulus intensity was 10 ma above sensory threshold; when the stimulus was only 1.5 ma above threshold, R_2 was much greater than R_1 at the same interstimulus interval. In the instance of Figure 31, conditioning and test stimulus intensities were equal. When conditioning stimulus intensity is varied and test stimulus intensity is kept constant, R_2 amplitude normally is greater with weaker than with stronger conditioning stimulus intensities (Shagass *et al.,* 1969).

Extracerebral Contaminants

The cortical origin of averaged scalp-recorded somatosensory responses seems to be supported by the contralateral spatial distribution, the variation in peak amplitude location in accordance with the area of the body stimulated, the relation to variations in stimulus intensity, and the recovery function results. More direct confirmation of cerebral origin comes from investigations in which responses have been recorded with electrodes placed directly on brain tissue as in the study of Libet *et al.* (1967) (Figure 14). Workers who have made observations of this sort include Jasper *et al.* (1960), Giblin (1964), and Domino *et al.* (1964, 1965). Figure 32 shows responses that were recorded simultaneously from scalp and dura, by Domino *et al.* (1965), during cryogenic surgery for Parkinsonism. Although

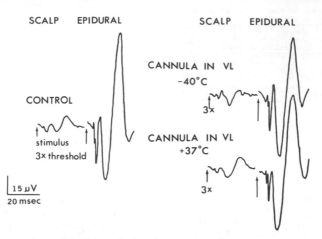

SCALP   EPIDURAL   SCALP   EPIDURAL

CANNULA IN VL
−40°C

CONTROL

3x

stimulus
3x threshold

CANNULA IN VL
+37°C

15 µV
20 msec

3x

NO MARKED EFFECTS OF VL CRYOGENIC LESION

ON SOMESTHETIC EVOKED RESPONSES

Fig. 32. Somatosensory responses recorded from the somesthetic area of the cerebral cortex from scalp needles and dura at operation. A left cryothalamectomy was carried out with two overlapping lesions in nucleus ventero-lateralis (VL). Placement of these lesions resulted in complete disappearance of the patient's rigidity and tremor with no sensory changes during or subsequent to the recording. With cannula in VL at -40 C, the evoked potential was approximately the same as it was before. Although greatly attenuated, the scalp recorded potential has the same shape as that recorded from the dura. (Reprinted with permission: Domino *et al.*, 1965.)

the scalp responses are markedly attenuated in comparison with those from the dura, the response shapes seem to be generally similar.

Most workers have found that averaged scalp responses are very much like those from dura or cortex, but this does not rule out the possibility that potentials of extracerebral origin can contaminate the averaged evoked response. The main extracerebral sources are the structures in the orbit and the head and neck muscles. The retinal–corneal potential is a major source of disturbance in recordings taken from the anterior portion of the head. Eyeblinks or eye movements give rise to potentials often exceeding 1000µv in amplitude from electrodes placed on or near the lid. Recordings of eye movements are designated the electro-oculogram (EOG). Another source of potential is the electroretinogram (ERG). These potentials are not usually troublesome in somatosensory recordings but their possible contribution to the late slow events, such as component 5 of Goff *et al.* (1962), cannot be discounted. In attempting to study the influence of transcranial polarization upon evoked responses, we observed changes which were entirely attributable to polarizing effects upon the orbital potentials (Ennever *et al.*, 1967). The danger of contamination may be increased if a reference lead is placed on the bridge of the nose or in some other location close to the eyes. Possible contamination by orbital responses can perhaps best be handled by making records from electrodes placed in the orbital region simultaneously with other averaged responses.

Myogenic contaminants of somatosensory responses arising under various conditions have been demonstrated by Cracco and Bickford (1968). Figure 33 (overleaf) gives an example of a response, peaking at about 50 msec after median nerve stimulation, induced by tension in the muscles of mastication. We have made similar observations, in our laboratory, in a few subjects, but were not able to do so regularly. The muscle response with biting tended to be equal bilaterally and its peak occurred more anteriorly than the peak of the somatosensory response. With neck tension, Cracco and Bickford found that there was phase reversal of the response at the inion. The true somatosensory response was little affected by muscle tension maneuvers, but some muscle responses had latency and distribution characteristics like those of somatosensory responses. In our experience, the usual effect of high-level muscle tension in a subject is the obliteration of regular evoked-response wave forms by the muscle "noise." Our failures to demonstrate "somatomotor" responses of the type shown in Figure 33 occurred mainly in psychiatric patient subjects; it may be that such patients find it difficult to maintain the sustained tension required to elicit the reflex. However, it seems clear that myogenic potentials represent a hazard that needs consideration in somatosensory response recording. Cracco and Bickford's main criterion, increase of amplitude with increasing muscle tension, would seem to be a valuable test of the validity of an evoked-response peak.

AUDITORY RESPONSES

The responses evoked by loud clicks and recorded from one electrode placed on the vertex and the other either in the occipital region, the mastoid, or the ear, vary in complexity according to different authors. Rapin (1964) described a series of eight early components occurring in both adults and children with the following mean latencies: positive, 8 msec; negative, 11 msec; positive, 17 msec; negative, 27 msec; positive, 38 msec; negative, 52 msec; positive, 56 msec; negative, 69 msec. She also described five late components with mean latencies as follows: positive, 80 msec; negative, 108 msec; positive, 121 msec; negative, 151 msec; positive, 164 msec. Rapin pointed out that the early components were recorded from electrodes placed on the posterior neck muscles as well as on the scalp, that they could be amplified by contraction and attenuated by relaxation of these muscles, and were reduced by drowsiness. These characteristics were considered by Rapin to confirm the view of Bickford et al. (1964) that the early waves result from reflex muscle contraction in response to the sound. The probable myogenic origin of the early auditory response components is generally accepted (Bergamini and Bergamasco, 1967; Davis et al., 1966). The careful

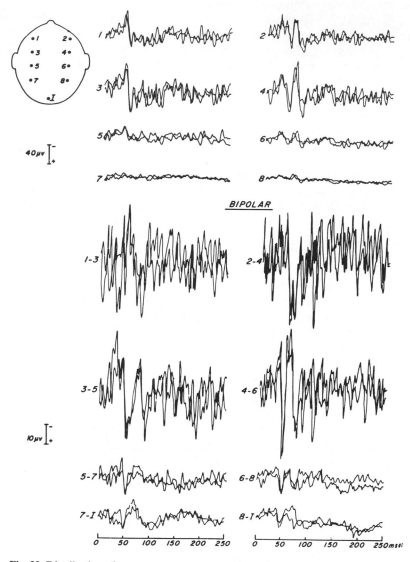

Fig. 33. Distribution of somatomotor response with tension in muscles of mastication. In nose-reference recordings, myogenic response is maximal in amplitude over motor region of the scalp (electrodes 3 and 4) on both sides. In bipolar recordings, this response undergoes gradient reversal over the scalp regions. (Reprinted with permission from Cracco and Bickford, 1968.)

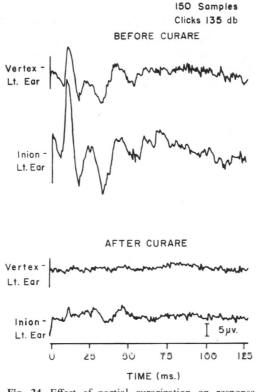

150 Samples
Clicks 135 db

BEFORE CURARE

Vertex –
Lt. Ear

Inion –
Lt. Ear

AFTER CURARE

Vertex –
Lt. Ear

Inion –
Lt. Ear

5 μv.

0   25   50   75   100   125

TIME (ms.)

Fig. 34. Effect of partial curarization on response evoked by intense click. After curare (sufficient to necessitate assisted respiration), response suppression was complete at the vertex, and almost complete at the inion. No anesthesia. (Reprinted with permission from Bickford *et al.*, 1964.)

localization studies of Goff *et al.* (1969) also indicate that, whereas the later components of the auditory responses are recorded maximally at the vertex in a consistent manner from subject to subject and are probably neurogenic, the earlier components appear to have a very inconstant distribution and are probably myogenic in origin. A more dramatic demonstration of the probable muscle source of these response components was provided by Bickford *et al.* (1964). As Figure 34 shows, the large early components of the response to very intense clicks were abolished when the subject was partially curarized to an extent sufficient to necessitate assisted respiration.

Myogenic contaminants do not present the only problem in recording auditory responses. Vaughan (1969) has demonstrated that the location of the auditory cortex in Heschl's gyrus probably results in a dipole orientation parallel to the surface of the scalp, so that the evoked response would be op-posite in polarity above and below the plane of the Sylvian fissure. This would explain the fact that the auditory response is about zero in a plane passing through the Sylvian fissure, where the primary cortex is located, and is maximal at the vertex.

Since the neural significance of the early components of the auditory response has appeared to be questionable, investigators have directed attention to the later components. Figure 35 gives some indication of the typical pattern and some variants of the slow potential evoked by tone pips as described by Davis *et al.* (1966). The most prominent components are the N_1 and P_2 deflections with latencies from 95 to 105 and 170 to 200 msec, respectively. The latency of P_1 ranges from 50 to 60 msec, and that of N_2 is about 300 msec. According to Davis *et al.* (1966), the latencies of the waves do not vary in relation to intensity of stimulus, except very near threshold, where they may be significantly prolonged. With tone pips, it

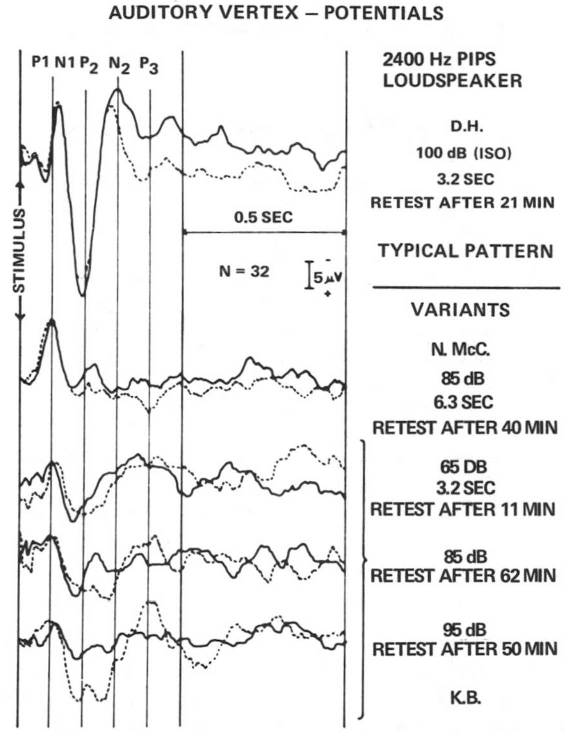

Fig. 35. Typical pattern and variants of slow evoked response to auditory stimuli. Upward deflection indicates relative negativity at vertex. 2400 Hz tone pips were presented by loudspeaker at regular intervals. Note variability of pattern in subject K.B. (Reprinted with permission from Davis *et al.*, 1966.)

also appears that latency is not a function of stimulus frequency between 300 and 4800 Hz.

Davis *et al.* (1966) studied the recovery of the component measured between N_1 and P_2. Their data suggest that the intervals between stimuli must be over 6 sec, and probably at least 10 sec, for maximal amplitude. With shorter interstimulus intervals, the amplitude of responses is reduced, e.g., at 1 sec the average amplitude is about one-quarter maximal. The results concerning the relation of amplitude to repetition interval were confirmed by Ritter *et al.* (1968). They found that, when tones were delivered every 2 sec, the amplitude of the vertex P_2 component dropped rapidly during the first few stimuli; when tones were delivered every 10 sec, amplitude remained steady. They interpreted the rapid decrease of P_2, with the faster rate of stimulation, as evidence of refractoriness within the auditory system. On the other hand, the amplitude of P_3, which occurs after 300 msec, appeared to be related more to the subject's state of attention than to stimulation parameters.

The consistency of the auditory response wave shape for a given individual appears to be high. Callaway *et al.* (1965) have used the correlation between successive data points of two averages as an indicator, and the coefficients normally exceed 0.90.

Possible Brain Stem Auditory Responses

Recently, Jewett *et al.* (1970) have reported detection of a very early response to click in scalp recordings. Employing a band-pass filter with lower cutoffs at either 10 or 300 Hz and a high cutoff at 2 kHz, and summing the responses to 2000 clicks, they demonstrated a series of low-amplitude waves which occur between 2 and 7 msec after the arrival of the stimulus at the ear. Although small (0.5 to 1.5μv), these waves were very consistent both within and between subjects in recordings from the vertex to the neck. Figure 36 shows examples of these early responses to click recorded in our laboratory; our results confirm those of Jewett *et al.*

Jewett *et al.* showed that stimulus reversal produced no reversal of polarity of the evoked potential; this indicates that no part of the cochlear microphonic was detected. They also showed that only the first negative wave (peak latency, about 2.4 msec) was present in recordings between the neck and wrist, indicating that the contribution of the neck reference to the wave form was minimal. Possible origin of the potentials in the middle ear muscles appears to be ruled out by the fact that middle-ear muscle potentials have a latency of about 10 msec with intense stimulation. Jewett *et al.* also found that when click intensity was reduced, the response showed an increased latency, change in wave shape, and reduced amplitude of the early waves. They suggested that the earliest electrical event recorded by them may originate in the cochlear nucleus.

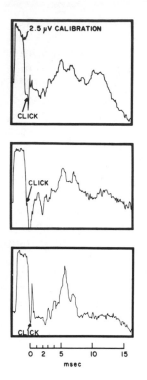

Fig. 36. Early auditory responses from three different subjects. Average of 2000 clicks regularly presented at 5/sec. Upper and lower filter settings 10 and 3000 Hz, respectively. Note relative consistency of pattern from subject to subject.

This auditory response seems to offer the possibility of studying very early central sensory events in conjunction with later cortical acitivity. However, simultaneous study of both early and late auditory events requires two time bases. The recording problem is similar to the one encountered in measuring peripheral nerve and cortical somatosensory responses simultaneously (Shagass and Schwartz, 1964a).

VISUAL RESPONSES

Flash stimuli of brief duration, and often of high intensity, have been favored by investigators. Figure 37 shows a series of schematic representations, compiled by Bergamini and Bergamasco (1967), of the configuration of the response as described by a number of authors (Gastaut *et al.,* 1963; Cigánek, 1961a; Barlow, 1957; Brazier, 1958; Contamin and Cathala, 1961; Cobb and Dawson, 1960; Schwartz and Shagass, 1964a; Van Balen, 1960; Van Hof, 1960; Rietveld, 1963; Kooi and Bagchi, 1964a). Although the lead derivations employed by the workers represented in Figure 37 were not uniform, and this would account for variations in components, there appears to be reasonable consistency between the findings.

The description of evoked-response wave shape that has been employed most often by other workers was provided by Cigánek (1961a). Figure 38 shows Cigánek's schematic diagram of the average response recorded by him with bipolar electrodes placed on the midline in the parietal and occipital regions. The response consists of a series of waves. Cigánek

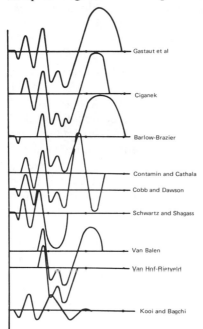

Fig. 37. Comparison of schematic diagrams of visual evoked potentials obtained by different investigators. (Reprinted with permission from Bergamini and Bergamasco, 1967.)

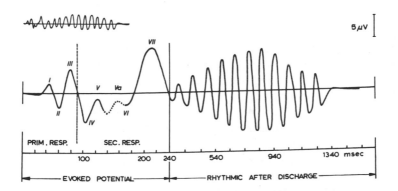

Fig. 38. Schema for the human response to light flashes recorded from midline occipital–parietal electrodes. Negativity at electrode Oz results in an upward deflection. Another time scale is used after 240 msec for the rhythmic after-discharge. At upper left the scheme for the evoked potential with a uniform time base is reproduced. (Reprinted with permission from Cigánek, 1961a.)

designated their peaks by successive numerals. Latency of onset of the response averaged 28 msec. The initial peak was negative with mean latency of about 40 msec. Mean latencies of waves II–VII were, respectively, 53, 73, 94, 114, 135, and 190 msec. Wave V frequently contained several peaks, one of which is labelled Va in Figure 38. Following the initial complex of the seven major peaks in Cigánek's scheme, there is a rhythmic after-discharge which is plotted in Figure 38 on a different time scale. This after-discharge has been termed "ringing" by Walter (1962). Kooi and Bagchi (1964b) have also presented normative data for the visual response.

Within the complex of waves preceding the after-rhythm, Cigánek distinguished two major constituents. He considered that waves I to III were the primary response and that the portion following wave III was the secondary response. He argued that the primary response was that specifically evoked in area 17 and cited as evidence its short latency and that it changed little with increase in frequency of stimuli and with sleep. In contrast, the secondary response had a long latency, decreased or disappeared when stimulus frequency was increased, and appeared to be nonspecific in the sense that it could be evoked by sound during sleep, and was affected by sleep. Cigánek, therefore, considered the secondary part of the response as being produced by nonspecific diffuse pathways. There is, however, considerable reason to doubt that even the earliest three waves of Cigánek's response originate in the primary visual area. Bergamini and Bergamasco (1967) conclude that the only true primary visual response would be the very early positive one described by Cobb and Dawson (1960), which has a latency of 20 to 25 msec, an amplitude of 1 to 1.5 μv, and is difficult to record from the scalp.

Rhythmic Afteractivity

The late afteractivity may be a very prominent feature of the response to a flash of light (Figure 38). According to Cohn (1964), it is seen in over 80 % of subjects with eyes shut, but in only about 20 % when the lids are open. Cohn also observed that, with a large number of flash stimuli, i.e., 300 or more in an averaging sequence, the amplitude of the after-rhythm became asymmetrical; it appeared to increase linearly with a number of flashes over only one occipital region, usually the right. The rhythmic afteractivity has its onset between 200 and 300 msec after the flash and may continue for an additional 1 to 1.5 sec. The duration of individual waves may range from 60 to 120 msec. The cerebral origin of the activity has not been disputed since it can be recorded from electrodes placed directly in the brain (Walter, 1962). Furthermore, it appears to be related to EEG alpha activity. However, the resemblance between the rhythmic after-activity and "spontaneous" alpha rhythm in form and frequency has led some investigators to believe that these two kinds of activity may be identical, and to attempt to verify this view.

Cigánek (1958) considered the after-rhythm frequency to be equal to that of the alpha rhythm, but Barlow (1960), using more accurate instruments, found that although the frequencies were close they were not identical. Cohn (1964) noted that the after-rhythm was dependent on the presence of spontaneous rhythmic EEG activity around the alpha frequency band. Our experience agrees with Cohn's. For example, in a study of patients with chronic brain syndromes, few of whom had EEG's with regular alpha activity, the most prominent finding was the absence of after-rhythm in response to light flash (Straumanis *et al.*, 1965). In view of the apparent close relation between alpha rhythm and the visual after-rhythm, Cohn (1964) proposed an explanation for the latter based on the effects of the light flash on the ongoing alpha activity. He suggested that the incoming flash first terminates the alpha activity and that then, following the flash, there is an interval during which the alpha frequency waves are reinstated. The light would thus set the phase of the alpha frequency, in a sense triggering new alpha activity after a delay. The alpha rhythm thus triggered would be roughly time-locked to the flash and be seen as the after-rhythm.

Cohn's view, that the after-rhythm is essentially alpha activity, is discordant with the observations of Walter (1962) in patients with implanted electrodes. He found that the after-rhythm was augmented by attention and by inhalation of CO_2, and attenuated by hyperventilation. The alpha rhythm reacts in an opposite manner to such manipulations. Furthermore, Perez-Borja *et al.* (1962) found that evoked responses to flash, sometimes followed by after-rhythms, were restricted to discrete areas near the calcarine region in implanted patients; in contrast, rhythmic activity, including alpha, was recorded from widespread areas. The issue of the identity between the rhythmic afteractivity and EEG alpha activity does not yet appear to be definitely settled, although there is agreement that they are probably intimately related. Creutzfeldt *et al.* (1966*a*) consider that they share similar mechanisms, namely the rhythmic synchronized and postsynaptic changes of cortical cell membranes.

Spatial Distribution and Extracerebral Contaminants

There is general agreement that at least the early components of the visual response originate in the occipital region. Cobb and Dawson (1960) demonstrated that the events of the first 90 msec after the flash reversed in phase between 3 and 6 cm above the inion. Vaughan's (1969) field distribution studies showed that both the early and late components were maximally recorded from electrodes overlying the occiput. He found that the later wave also possessed a secondary maximum in the central region, but, on further scrutiny, it appeared that the central and occipital components were not the same; he decided that they probably represented two distinct generators. Goff *et al.* (1969) considered their distribution data for the visual response to be less clear than those for the auditory and somatosensory systems. They found that there was considerable variability in the distributions of components both as to extent and focus and that some showed multiple foci. They concluded that an electrode located in the occipital area would be optimal for recording early components, but that a vertex electrode would be desirable for picking up the nonspecific later potential. Our own attempts to define the locus of origin of visual evoked-response peaks by seeking phase reversals in recordings from a bipolar chain of electrodes have not yielded satisfactory results. Not only are foci of a given peak variable between subjects, but it appears that different peaks have different foci. Apparently, potential gradient studies employing a common reference lead, such as one on the contralateral ear lobe, may yield more uniform results, although the comments of Goff *et al.* (1969) suggest that the difficulties are not eliminated by this approach.

　　　In some subjects, the bipolar chain recording method does reveal a clear point of reversal for all components. Tracings from one subject, a patient with brain syndrome and myoclonic facial twitches, are shown in Figure 39. The point of reversal was at electrode 6 in the midline, 9 cm above the inion. Careful examination of the records in Figure 39 will also reveal an early small deflection, with a peak latency of about 30 msec, which reverses in phase at electrode number 2, 3 cm below the inion; this appears to be a neck muscle response. It occurs earlier than the first facial twitch response to flash (electrodes 15–16), which had a peak latency of 43 msec. Figure 39 thus demonstrates a clear midline parietal focus for the cerebral response, another focus for the neck response of latency similar to the earliest visual component, and different timing for the facial myoclonic twitch. It illustrates the importance of recording simultaneously from many electrodes.

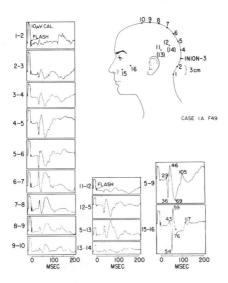

Fig. 39. Localization by phase reversal of response evoked by flash in patient with brain syndrome and myoclonic twitches of face. Entire early portion of response reverses at electrode 6. Center tracings show that reversal is in midline (electrodes 13 and 14 were located on right side in positions equivalent to 12 and 11, respectively). In addition to phase reversal at lead 6 in midline, amplitudes diminish anterior to lead 8 and posterior to lead 4. Tracings at right compare midline head recordings with those on face and illustrate latency differences (numbers are peak latencies in msec) between cerebral and myogenic response. Upward deflection indicates relative positivity of posterior or left-sided lead of pair. (Reprinted with permission from Shagass, 1968*b*.)

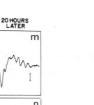

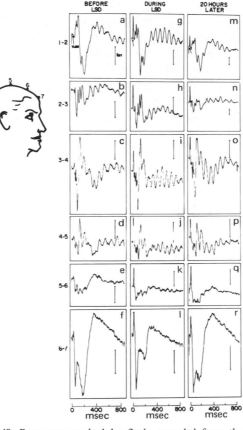

Fig. 40. Responses evoked by flash, recorded from the chain of bipolar midline electrodes. Note large potential recorded from anterior electrodes; this contains activity generated in ocular structures. While subject was under influence of LSD, the after-rhythms recorded from leads 1–2 and 3–4 were of larger amplitude and faster frequency than those obtained before LSD or 20 hr later. (From Shagass, 1967a.)

When responses to flash are recorded from electrodes placed near the eyes, an extremely large potential originating in orbital structures is recorded (leads 6–7, Figure 40). This contains both the ERG and EOG. In our work, this potential was more pronounced with flash than with other stimuli. Recording from electrodes anterior to the vertex will almost certainly reflect potentials of orbital, in addition to cerebral, origin. Additional extracerebral potentials can be recorded posteriorly. Bickford *et al.* (1964) demonstrated that very large muscle responses can arise from maneuvers causing tension in the muscles of the neck. The extent to which such myogenic potentials are recorded under ordinary conditions is uncertain. There is some evidence that they may not be too important. For example, Domino and Corssen (1964) were able to show that paralysis of the skeletal musculature with succinylcholine did not produce any marked alterations in visually evoked responses recorded in patients studied before scheduled surgical operations. Furthermore, Katzman (1964a) concluded from a summary of various kinds of evidence, including failure to modify visual responses by neck contraction and observations of similarity between scalp and direct cortical recorded responses, that the usual averaged visual responses are most likely of cerebral origin. It should, nevertheless, be clear that the hazard of myogenic contamination cannot be overlooked.

According to Barlow (1964b) and Gastaut and Regis (1965), the positive wave of the visual evoked response with 100 to 120 msec latency (Cigánek's wave IV, Figure 38) is identical with EEG lambda waves. Lambda are random, simple sawtooth waves, positive at the occiput, which can be observed in the EEG when the eyes are open, and more regularly when there is a sharp change in the visual field (Evans, 1953). Vaughan (1969) has drawn attention to the fact that an evoked potential corresponding to the lambda wave appears in the occipital area following each eye movement. It also appears that the configuration of Cigánek's wave IV is dependent on the relative predominance of scotopic and photopic vision, as indicated by changes with different levels of dark adaptation (Gastaut et al., 1966).

Reliability of the Response to Flash

Although all workers have noted considerable variability of the response to flash between subjects, and in the same subject from time to time, the degree of stability for a given subject under similar conditions appears to be quite great. Kooi and Bagchi (1964b) reported test–retest correlations ranging from 0.87 to 0.97 for measurements made during the same test session. The issue of long-term stability was studied by Dustman and Beck (1963). They recorded responses from seven male adult subjects at intervals of 1 week or more. The similarity between responses for a given subject was determined by computing correlations between 25 equally spaced measurements during the 300 msec following flash. The median coefficient was 0.88. Similar correlations were computed between responses of different subjects; the median of these was only 0.37. In addition to demonstrating high reliability, Dustman and Beck's data provided support for the conclusion that the visual evoked response appears to be relatively unique for an individual. In another study, Dustman and Beck (1965) compared monozygotic and dizygotic twins, triplets, and unrelated children. The mean correlation for the monozygotic twins was 0.82; for the dizygotic twins, it was 0.58. Some correlations were higher between members of a twin pair than between the responses of the same twin taken some time apart.

It would appear that the flash evoked response is generally stable and relatively characteristic for an individual, although some persons, according to Dustman and Beck (1965a), are more variable than others. These findings concerning reliability are particularly important for the kind of psychiatric research that attempts to relate evoked potential and personality characteristics. Lack of stability in the physiological measure would discourage the possibility of personality correlates.

Effects of Stimulus Intensity

Cobb and Dawson (1960) noted that light flashes of reduced brightness gave responses of smaller amplitude and more prolonged latency. Systematic

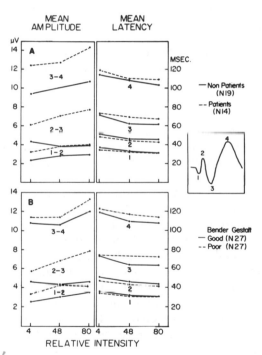

Fig. 41. *A*, intensity–response curves for amplitude and latency of visual responses in patient and nonpatient groups. *B*, curves for 54 subjects (including 43 patients) divided with respect to Bender Gestalt performance. Note similarity between patient –nonpatient differences and those related to Bender performance. Amplitudes tend to increase and latencies to decrease with increasing intensity of light flash. (Reprinted with permission from Shagass *et al.*, 1965.)

data leading to a similar conclusion were presented by Diamond (1964), and most other workers agree with Cobb and Dawson's findings. An example from our own data is given in Figure 41. This shows the relationship between stimulus intensity and measurements of several amplitudes and latencies in a group of subjects divided according to psychiatric and psychological test criteria (Shagass *et al.,* 1965). The inverse relationships between stimulus intensity and latency and the positive ones between intensity and amplitude are demonstrated. In our studies we noted a number of exceptions to the general trend, i.e., we found subjects who showed *smaller* responses with brighter flash. A possible explanation for this paradox now seems to be available from the work of Buchsbaum and Silverman (1968), who have demonstrated that the intensity–response curve is related to a personality factor involved in stimulus intensity control. Their work will be considered later.

Recovery Functions

Although the pioneer studies of visual recovery functions by Gastaut *et al.* (1951*a*) initiated our interest in evoked-response research, recovery function studies of averaged visual responses have proved to be difficult. This is because the primary component is small and hard to record, the remainder of the response is very complex, and many factors, such as eye position, are hard to control. Because of these problems, we have conducted few systematic studies of visual recovery functions. Figure 42 gives sample tracings obtained in one of these (Shagass and Schwartz, 1965*a*). It will be seen that, with all interstimulus intervals, shape of R_2 differs considerably from that of R_1. Also, a number of additional components appear to be introduced in some R_2's, particularly with intervals under 100 msec. Figure 43 shows mean latency and amplitude recovery curves for 60 subjects; measurements were made for 23 interstimulus intervals, the longest being 200 msec. It took nearly 200 msec for latencies of R_2 to diminish and approach the latencies of R_1. In contrast, amplitude recovery appeared to start before 20 msec, and the maximum R_2 amplitude of component 1–2 was approached by 30 msec and that for component 2–3 by 90 msec. R_2 of the later component measured (3–4) did not reach maximum amplitude until 120 msec. Amplitude recovery appeared to take place more rapidly than latency recovery.

Stimuli Other Than Flash

Our discussion of the visual response, has, to this point, dealt mainly with the effects of flash stimulation. Major differences in evoked-response characteristics are observed when other types of visual stimuli are employed. Figure 44 gives an example of the response wave forms obtained with sine-wave-modulated light (Callaway, 1968). The response to this form of light is roughly sinusoidal. This sinusoidal response seems similar to the

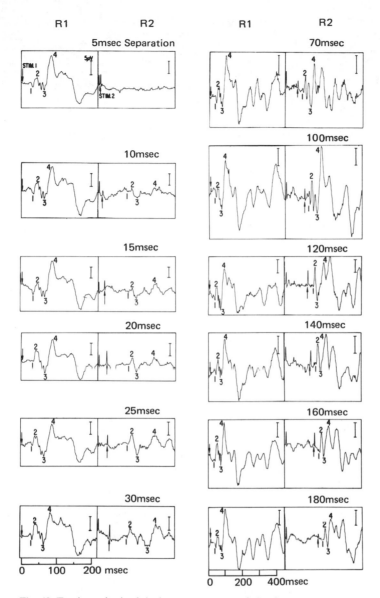

Fig. 42. Tracings obtained during measurement of visual response recovery functions in one subject. Upward deflection indicates positivity at occipital electrode. R_2 tracings are average of 50 $(R_1 + R_2)$ minus 50 R_1. Measured points numbered from 1 to 4. (Reprinted with permission from Shagass and Schwartz, 1965a.)

one noted by Cigánek (1961b) with light flashes of 10 to 16 per second, so that the waveform may depend less upon sine-wave modulation than upon stimulus frequency.

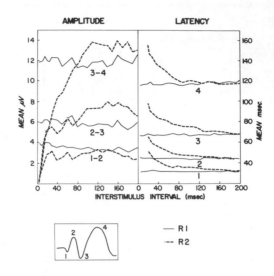

Fig. 43. Mean amplitude and latency recovery curves of visual responses for 60 subjects. Peak designations are the same as those in Fig. 42. (Reprinted with permission from Shagass and Schwartz, 1965a.)

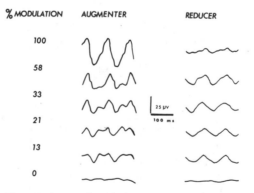

Fig. 44. Averaged evoked responses to sine-wave-modulated light. Note regular increase of amplitude in augmentor as percentage modulation is increased. Similar increases absent in reducer subject. (Reprinted with permission from Callaway, 1968.)

The form of visual evoked responses may be quite different with patterned than with unpatterned light. Spehlmann (1965) found that responses to patterned light contained a positive late wave with a latency of 180–250 msec which was localized in the occipital region. Figure 45 illustrates the effect of light patterning on the configuration of the evoked response. It appears that the critical factor is the density of contrast borders between black and white lines of the stimulus pattern. Lifshitz (1966) has also demonstrated that slides containing pictures of different emotional value, ranging from neutral to erotic, resulted in evoked-response patterns which differed measureably from responses to the same slides made non-associational by defocusing or to blank light flashes. Clynes *et al.* (1967) have studied the effects of variations in pattern and color. The tracings in Figure 46 were made from pairs of electrodes placed over the left occiput, in a small circle ("rosette"), with simultaneous recordings taken from opposite pairs of electrodes representing angles of 0, 45, 90, and 135 degrees, respectively. This array permits components to be recognized as of different spatial origin if they peak in different traces. The records in Figure 46 illustrate differences related to pattern, to degree of focus, and to color pattern. The data of Shipley *et al.* (1965) corroborate the color differences and suggest an interesting application; they found that visual response variations with color were absent in a color-blind observer. Beatty and Uttal (1968) found a systematic relationship between the amplitude of a visual evoked response component at about 100 msec and the degree of grouping of stimuli presented on a cathode-ray screen; amplitude was smaller with the grouped stimuli.

The relationship between visual evoked responses and stimulus

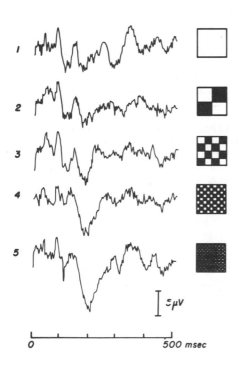

Fig. 45. Responses to visual patterns of increasing contrast density in one subject. Each tracing is average of 100 responses. The patterns used for tracings 2 to 5 reflect the same amount of light, but the number of contrast borders on them increases in the relation of 1:2:16:32, as indicated in the sketches of the patterns on the right. Tracings were obtained in irregular sequence during the experiment. Note increases of amplitude and latency to the peak of the positive (downgoing) component at about 200 msec. (Reprinted with permission from Spehlmann, 1965.)

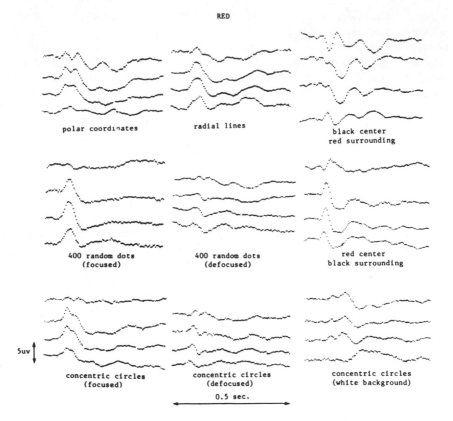

Fig. 46. Examples of varieties of space-time patterns obtained from various visual field structures of lines and shapes. Lines are projected on black background except at bottom right. Amplitudes of responses drop sharply when images are defocused. Color differentiation is marked for the defocused images, but this seems masked by the edge sensitive responses to the focused images. Top right is a response to a circular, black central field, 12 inches in diameter, surrounded by red. Below it is a red central field of similar size, surrounded by black. With these response shapes, component analysis reveals the existence of four main independent spatial components of different latency. Responses to radial lines and circles appear to be basic. Bottom-right pattern illustrates the observation that the responses to lines are very sensitive to the color and intensity of the surrounding field. The variation with different visual stimulus conditions is striking. (Reprinted with the permission of Clynes *et al.,* 1967.)

attributes is clearly a topic of major importance, but a detailed description is beyond our scope. The examples that have been given should suffice to draw attention to the fact that any change in the form, pattern, color, or meaning of stimuli may alter the evoked response.

MOTOR POTENTIALS

Bates (1951) was apparently the first to describe brain potential correlates of movement in man. Using the technique of superimposition, he was able to discern a cortical potential beginning 20 to 40 msec after the onset of contraction. Kornhüber and Deecke (1965) described a "readiness potential" *(Bereitschaftspotential)*, which was enhanced by "intentional engagement" of the subject. Gilden *et al.* (1966) employed averaging to demonstrate the existence of additional aspects of the cortical potential associated with voluntary movement. Their technique involved placing a delayed trigger signal on one track of the magnetic tape on which EEG and EMG were recorded. Analysis of the data was accomplished by playing the tape backwards so that the delayed trigger signal could be used to start the action of the computer to summate EEG and EMG activity.

Figure 47 illustrates the motor potential associated with dorsiflexion of the foot. The main components are as follows: (1) a slow negative shift of the base line starting about 1 sec prior to onset of EMG evidence of muscular contraction (the EMG line in the figure represents the summation of full-wave rectified EMG); (2) about 50 to 150 msec prior to contraction, the slow negative variation terminates in a small positive deflection

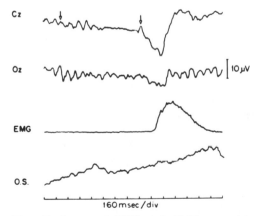

Fig. 47. Summated EEG and EMG potentials associated with dorsiflexion of the left foot. The onset of the contraction is preceded by a slow negative shift beginning at the point indicated by the left arrow and also by a positive–negative deflection indicated by the second arrow. Scalp negativity gives downward deflection. EMG recorded from anterior tibial muscle. OS: activity from electrode below left eye. (Reprinted with permission from Gilden *et al.,* 1966.)

shown by the second arrow over the *Cz* line; (3) this is followed by a larger negative wave which develops a peak amplitude of 10–15 µv during the rising phase of the summated EMG; (4) immediately following the negative deflection, and 50–150 msec after contraction began, a larger positive wave, 20–30 µv in amplitude, is recorded. Gilden *et al.* (1966) found that there was general reproducibility of the motor potential for a given movement in a given subject. The motor potentials following fist contraction, although generally similar in form to those associated with foot dorsiflexion, were more variable from subject to subject. Ertl and Schafer (1967) have demonstrated averaged waveforms preceding a single repetitively spoken word; these seem to be motor potentials. However, they did not control for the possibility that they may have summed activity arising in the tongue musculature.

Vaughan *et al.* (1968) conducted a careful study of the topography of the motor potential. They examined its field distribution for movements in various body areas and compared these with the distribution of somatosensory responses and of the retinal–corneal potential. They concluded that the motor potential is maximal over the Rolandic cortex and that its somatotopic distribution for contractions of various muscles was similar to that obtained by direct cortical stimulation at neurosurgery. They found that the field of the motor potential was compatible with its generation by a source of size comparable to that of the excitable motor cortex. The late component of the somatosensory evoked response had a distribution similar to that of the motor potential, but it was extended slightly posterior.

Gilden *et al.* (1966) noted that the appearance of the initial long negativity in the motor potential resembled that described for the contingent negative variation (CNV) (Walter *et al.*, 1964*a*). However, the distribution of the motor potential seemed different from that described for CNV. Vaughan *et al.* (1968) found that if they introduced a fixation field to ensure ocular stability, and thereby markedly reduced contamination by EOG, the negative potential showed a maximum in the Rolandic area. Figure 48 shows the motor potentials associated with dorsiflexion of the right foot, as recorded from several adjacent electrodes each referenced against the linked ears. It demonstrates that the point of maximum amplitude was found at the middle electrode of the three placed in a line anterior to the postulated Rolandic fissure.

Gilden *et al.* (1966) and Vaughan *et al.* (1968) consider the slow negative component of the motor potential to be associated with processes occurring in preparation for movement. The abrupt negative or positive–negative deflection occurring just prior to and during muscle contraction was thought to perhaps reflect activation of pyramidal tract neurons. The late positive component of the motor potential was attributed in part to afferent feedback generated by movement.

Motor potentials have not so far received much attention in psychiatrically oriented studies. However, as an objective indicator of cortical ac-

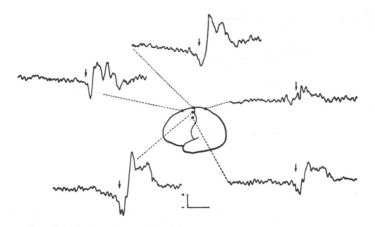

Fig. 48. Motor potentials associated with dorsiflexion of the right foot. Sum of 400 contractions. Calibration, 500 msec and 2.5 μv. (Reprinted with permission from Vaughan *et al.*, 1968.)

tivity related to movement, it seems likely that this potential will generate increasing interest. It may, for example, be used in attempts to measure intracortical delay time, i.e., the time elapsing between arrival of sensory impulses at the cortex and initiation of acts signaled by the sensory stimuli. With averaging techniques, the transmission time to the cortex and the time of initiation of motor potential events preceding muscle contraction can be measured; from these, estimates of the intracortical delay interval are possible. Psychological correlates of such delay intervals seem worthy of study.

LONG LATENCY POTENTIALS

Vaughan (1969) allocates a separate category to the long latency (300–500 msec) positive components of responses which appear to be related more to complex psychological variables than to the physical attributes of the stimulus. These potentials were originally described as a correlate of stimulus uncertainty by Sutton *et al.* (1965). The activity may be elicited by the *absence* of an expected stimulus (Sutton *et al.*, 1967). Figure 49 illustrates the nature of the positive response, its elicitation by an experimental condition producing uncertainty as to whether a stimulus would be delivered, and the effect of expected stimulus interval upon the latency of the response. All of the average responses shown in Figure 49 were evoked by physically identical single clicks recorded between vertex and left ear lobe. In the first tracing (certain) the subject was informed, in advance, whether the click would be single or double. In the second tracing, the response is also to a single click, but the subject did not know whether

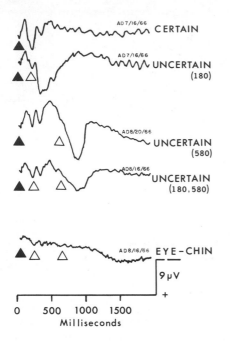

Fig. 49. Average responses to single clicks obtained from one subject under several experimental conditions. First triangle indicates actual delivery of click. Subsequent triangles indicate points in time when a second click might have occurred but did not. Note delay of P300 wave when a second click is anticipated at a later time. (Reprinted with permission from Sutton *et al.*, 1967.)

the click would be single or whether there would be a second click after 180 msec. In the third tracing, the alternative double click would have occurred with an interval of 580 msec after the first; the positive potential is seen to be shifted in latency. The fourth response shown in Figure 49 was obtained under conditions in which the stimuli could be either single click, a double click with a 180 msec interval, or a double click with a 580 msec interval. However, the subject was asked to guess only whether the forthcoming stimulus would be single or double. Since the absence of a click at 180 msec could still leave the possibility of a click at 580 msec, the positive process appears to be delayed until that time. Sutton *et al.* (1967) concluded that the large positive process was related to the resolution of uncertainty. This seems compatible with the views of Ritter *et al.* (1968), who demonstrated long-latency-positive responses whenever the occurrence of auditory stimuli was unpredictable. They concluded that the 300 msec component reflected a shift of attention associated with the orienting response.

CONTINGENT NEGATIVE VARIATION AND OTHER STEADY POTENTIAL SHIFTS

Köhler *et al.* (1952) showed that steady potential shifts occurred during prolonged auditory and visual stimuli. Their recordings from the human scalp were made without the use of averaging techniques. The initial aspect of the motor potential and the "readiness potential," described earlier, also consists of a negative steady potential shift. The phenomenon in this category that has received the greatest amount of attention is the contingent negative variation (CNV) or expectancy (E) wave described by Walter *et al.* (1964a). Cohen (1969) has recently provided an excellent survey of the literature on CNV.

The most commonly used situation for recording CNV employs a paradigm similar to that used in measuring reaction time. An alerting, or warning, signal precedes the delivery of an "imperative" signal, to which the subject must make some kind of response, such as a button press. The interval between stimuli is usually 1 to 2 sec; a minimum of 0.5 is required for adequate development of CNV (Walter, 1969). In the usual experiment, there is also a control condition in which a similar signal, without the same significance for responding, is administered. In the experiment giving the records shown in Figure 50, a warning click in the right ear signified that the response button was to be pressed after the flash, whereas a click in the left ear did not have this meaning. Under the "response" condition, as illustrated in Figure 50, there is a slow potential deviation, negative at the vertex, of the CNV, which commences about 200 msec after the stimulus. The CNV continues to rise until the imperative

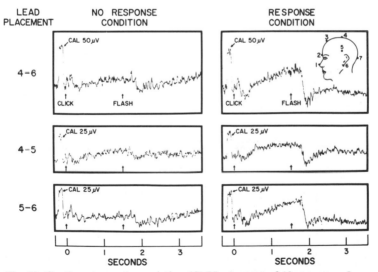

Fig. 50. Contingent negative variation (CNV). Average of 20 responses. In response condition, a click in the right ear signifies button press after the succeeding flash. In no response condition, left click signifies button not to be pressed. Upward deflection indicates relative negativity in upper electrode. Note slow potential shift in response condition. (Reprinted with permission from Straumanis *et al.*, 1969a.)

stimulus is delivered and responded to, at which time there is a precipitous shift back toward the base line or positivity. The "no response" condition did not elicit a CNV.

CNV may also be demonstrated independently of a motor response, as when a subject anticipates seeing a picture or is instructed to "think" a word after the imperative stimulus (Cohen, 1969).

The occurrence of a slow potential phenomenon of cerebral origin in association with various psychological conditions such as "expectancy," "conation" (Low et al., 1966), or "motivation" (Irwin et al., 1966) seems reasonably well established. Walter (1964b) obtained CNV from implanted subdural electrodes in man and Low et al. (1966) recorded it from implanted electrodes in monkeys during conditioning. However, as Figure 51 illustrates, the CNV recorded at the vertex may be very difficult to distinguish from concurrent EOG activity. The recordings in Figure 51 were made simultaneously with those in Figure 50. In the "response" condition, a

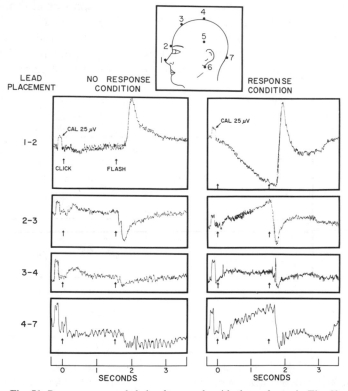

Fig. 51. Responses recorded simultaneously with those shown in Fig. 50. Upward deflection indicates relative negativity in anterior electrode. Note the large potential generated under the response condition from leads placed near the eye. (Reprinted with permission from Straumanis *et al.*, 1969a.)

very large potential shift was recorded from electrodes near the eye; this was not present during the "no response" condition. The ocular potential shift paralleled the CNV. In our study of CNV (Straumanis *et al.*, 1969*a*), we found high correlations between CNV effects recorded from leads placed near the eyes and those recorded between either vertex and mastoid or temporal area and mastoid (compare leads 1–2 in Fugure 51 with leads 4–6 and 5–6 in Figure 50). Thus, although we were able to demonstrate that, in some subjects, CNV occurred independently of EOG changes, this did not appear to be the case for most. The EOG changes seemed to be a concomitant of downward eye movement, accompanying the button press, since potential shifts resembling the CNV could be produced by voluntary movement of this sort. On the other hand, if the eyes were held steady by instructions to fixate a pattern, the EOG changes could be virtually eliminated.

While it seems clear that ocular fixation is a desirable condition for CNV recording, it is also essential to monitor the EOG. The problem of EOG contamination of CNV can be reduced by other strategies. The method suggested by Walter (1967) is to link a mastoid lead to a supra-orbital electrode through a potentiometer. The resistance between the two reference leads is adjusted so that the voltage brought about by eye movement is sufficient to cancel the EOG artifact at the vertex electrode. Another approach is to correct CNV measurements by means of a factor derived from the EOG recording (Hillyard, 1969*a*). These strategies do not compensate for horizontal eye movements, which are picked up particularly at the mastoid lead (Overton and Shagass, 1969). Also, since we have found that the distributions of eye movement and eyeblink potentials are not identical, a compensatory factor based on one of these will not adequately correct for the other.

Vaughan *et al.* (1968) have raised some question about the extent to which slow negative potentials can be considered independent of the slow negative phase of the motor potential, localized in the Rolandic area and maximal over the hemisphere contralateral to the movement, which seems identical with the "readiness" potential of Kornhüber and Deecke (1965). Although Low *et al.* (1966) found that CNV was greatest in the anterior frontal region, Walter (1967) and Cohen (1969) reported maximal amplitude at the vertex with both bipolar and monopolar derivations. Since electrode placements appropriate for determining the lateralization of the CNV have seldom been employed, it seems possible that the maximum activity at the vertex may reflect motor potential activity. Opposed to this are observations that CNV can be demonstrated without motor responses, but this cannot be taken as absolute proof since covert motor activity could be involved in thinking a word or expecting a picture. McAdam and Seales (1969) obtained data relevant to the issue by recording the "readiness" potential under reward and nonreward conditions. Under both conditions, the negative slow wave response was lateralized, with the amplitude on the hemisphere contralateral to the responding hand about twice as great as that seen ipsilaterally. However, the reward condition about doubled the amplitude of the "readiness" potential. It appeared, therefore, that motivation, a factor associated with CNV, increased the amplitude of the potential, but that lateralization was maintained. The results of McAdam and Seales suggest that the motor potential and CNV are both amenable to psychological manipulation. Although it remains to be established to what extent they are independent phenomena, they must vary together to some extent in the commonly employed experimental situations.

Cohen (1969) found that mean CNV amplitude at vertex was 21.4 μv (standard deviation about 4 μv) in 60 young adults and that the test–retest reliability was 0.8 in 34 subjects who were studied in the same situation in two sessions separated by 2 to 8 days. Straumanis *et al.* (1969*a*), on the other hand, found that when measurements were made in two different experimental conditions in the same subjects on the same day, there was very little consistency of CNV amplitude from one condition to another. These results underscore the importance of the situation as a determinant of CNV responsiveness.

While the CNV phenomenon has excited much interest (Dargent and Dongier, 1969) its application to studies of psychiatric patients requires caution. Avoidance of the misleading results that may be introduced by EOG contamination is a difficult problem. The most desirable control feature is ocular fixation, which psychiatric patients find particularly difficult to sustain. EOG can be monitored and correction attempted, but these correction procedures require assumptions about the distribution of eye movement that may not necessarily obtain. Nevertheless, there have been some apparently successful attempts to measure CNV in disturbed patients (Small and Small, 1969).

Chapter 4

Age, Sex, and Other Factors

Evoked-response recording has provided a new tool for studying maturation and aging of the central nervous system (CNS). There is also evidence that the method may yield interesting information concerning sex differences. The manner in which responses may vary with age and sex is of obvious methodological importance in psychiatric studies; experimental designs must take such variations into account. Handedness, time of day and cardiorespiratory cycles are other factors of methodological significance to be considered in this chapter.

SOMATOSENSORY RESPONSES AND AGE

Infancy and Childhood

Desmedt *et al.* (1967) and Manil *et al.* (1967) recorded somatosensory responses from normal newborn infants. Figure 52 shows that the neonatal responses differed considerably in wave shape from those of adults. The most prominent component was a large negative deflection with onset latency from 16 to 24 msec and duration from 12 to 18 msec. This was preceded by a surface positive component with latency from 12 to 18 msec which was more prominent in anterior leads. Desmedt *et al.* (1967) also measured conduction velocity in the neonate median nerve and compared it with that of the adult. The mean velocity of neonates was 30 m/sec, much slower than the 62 m/sec for adults. Taking into account the length of the pathway, Desmedt *et al.* concluded that the latency of somatosensory responses was relatively long in the neonate. Manil *et al.* (1967) found the amplitude of neonate responses to be similar to that of adults.

Laget *et al.* (1967) studied somatosensory responses in 40 children ranging in age from about one month to 14 years. Their data indicate that the large initial negative wave described by Desmedt *et al.* is predominant

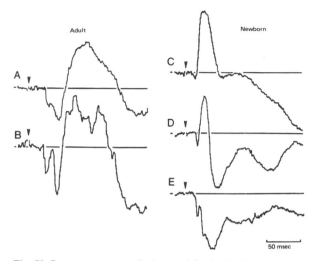

Fig. 52. Somatosensory evoked potentials obtained by summing 200 responses in two adult subjects (*A* and *B*), in an infant 7 hr old, weighing 3.4 kg *(C)*, and in a baby 18 hr old, weighing 3.3 kg (*D* and *E*). In *E*, the active electrode is placed in the anterior portion of the focal region. The electrical stimulus was applied to a finger of the contralateral hand in *A* and *B*, and over the median nerve at the wrist in *C* and *D*. Negativity in active electrode gives upward deflection. (Reprinted with permission from Desmedt *et al.*, 1967.)

until about 6 months of age, at which time the positive wave (peak 4 in our numbering scheme) becomes predominant. Laget *et al.* consider the adult wave form to be achieved at about the age of 4 years. In the young infant, the peak latency of the initial negative deflection is about 30 msec, but this decreases to about 20 msec at the age of 6 months. The amplitude of the somatosensory evoked response appears to be quite small during the first six months and then increases markedly. The results presented by Laget *et al.* suggest that peak latency of the initial deflection remains constant, in the vicinity of 20 msec, from the age of about 6 months to 14 years. This latency plateau indicates that there is a remarkable parallel between the increase in the length of the nervous conduction pathway and in the velocity of conduction. Laget *et al.* explained this as resulting from concomitant increases in the length, diameter and myelinization of the nerve fibers.

Hrbek *et al.* (1968) studied cerebral potentials evoked by stretching a muscle to elicit a monosynaptic reflex, employing a tendon tap as stimulus. They considered the evoked response to be proprioceptive. Most of the observations were made in neonates during sleep, and they found that the early and late components of the proprioceptive response differed according to sleep stage. During "regular" sleep a positive wave of more than 200 msec latency was much larger than it was during "irregular" sleep. Latencies of the responses in the infants were prolonged compared to those of adults but, disregarding the time scale differences, Hrbek *et al.* considered the wave shapes to be quite similar. The resemblance of the infant response form to that of adults led Hrbek *et al.* to conclude that the proprioceptive response is more mature in the neonate than is the visual response, which does not resemble that of adults in wave form. Since Hrbek *et al.* did not demonstrate the large initial negative component, their findings differ from those of Desmedt *et al.* (1967) and Laget *et al.* (1967). The differences may be due to the type of stimulus employed.

The available information indicates that somatosensory responses can be elicited at birth, that they are of prolonged latency, and that they begin to assume adult shape at about the age of 6 months or earlier.

Adolescent and Adult Years

We obtained data concerning the variation of somatosensory response characteristics with age in a study of recovery functions (Shagass, 1968a). In this study there were 19 interstimulus intervals ranging from 2.5 to 120 msec. Each R_1 represented the average response to 50 stimuli. The numerical scheme shown in Figure 24 was employed to designate ten peaks for measurement. For each subject the mean of the 19 measurements of each R_1 amplitude and latency was used in statistical analysis, with the exception of the values for peaks 2 and 3, which were often not detectable. As seen in Table II, the frequency with which peaks 2 and 3 were seen in normal subjects was a function of age. They were seldom observed in subjects 40 years of age or older.

The statistical significance of variations with age of the amplitude and latencies of the remaining eight peaks was assessed by dividing the

Table II. Incidence of Visible Peaks 2 and 3 in Somatosensory R_1 by Age

Age	Total number	Number 2 and 3	% 2 and 3
15–19	17	10	58.8
20–24	19	13	68.3
25–29	10	5	50.0
30–39	13	5	38.4
40–59	13	1	7.7
60+	17	1	5.9
Total	89	35	39.3

nonpatient subjects into four age groups for analysis of variance. Since simultaneous analyses were performed for age and sex, some subjects had to be discarded to achieve proportionality in the cells; this left 76 subjects out of the original 89. The mean amplitude and latency values are shown in Figure 53. The statistical tests revealed that the amplitude of peaks 1, 6, 7, and 9 varied significantly with age; with increasing age, peaks 1 and 9 became more negative, and peaks 6 and 7 became more positive. The only latency measures significantly related to age were those for peaks 1 and 10, both of which increased with advancing years. The statistical significance of changes in each variable occurring over the course of the 19 determinations was also assessed. R_1 amplitude 5 and latencies 5, 6, and 9 showed systematic shifts in the direction of decreased amplitude and longer latency. These changes suggest "habituation" over the course of the experimental session.

From the foregoing results it would appear that, in general, somatosensory response amplitudes tend to increase with age and that some latencies tend to become more prolonged. Figure 54a presents the findings in another form; it shows the composite mean evoked response for subjects over and under age 40.

A further statistical analysis was performed by multiple linear regression. In this analysis, the multiple correlation between all evoked response variables and age was determined for 86 subjects and the significance of beta coefficients in the regression equation was tested. The variable with the least significant beta coefficient by F test was then eliminated and the

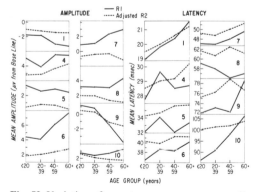

Fig. 53. Variation of somatosensory response amplitude and latency with age. R_1 measurements based on response to first stimulus of the pair in recovery-function determination. R_2 measurements based on response to second stimulus, adjusted for covariance with R_1. All measurements averaged across 19 interstimulus intervals from 2.5 to 120 msec. Numerical designation of peaks as in Fig. 24. For further details, see text.

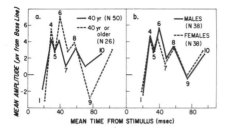

Fig. 54. Composite mean somatosensory evoked responses to single stimuli 10 ma above threshold for 76 nonpatient subjects divided according to age and sex.

computations were repeated without this variable. There were successive recomputations until all remaining beta coefficients were significant. The final yield was a regression equation with minimal redundancy of variables and the corresponding multiple coefficient (R). With this analysis, the initial R for 16 R_1 variables was 0.811, and the final R was 0.741 ($p < 0.001$). Only two of the 16 variables entered into the initial multiple regression remained at the end. These were the amplitude of peak 9 and the latency of peak 1. The optimal combination of these two somatosensory evoked-response measurements would therefore reduce from chance the error of predicting age by about 55 %. Later occurrence of peak 1 and greater negativity of peak 9 would be more likely in an older person. These results are in accord with the impression gained from inspection of Figure 54a, but some of the differences apparent in this figure were sufficiently highly correlated with latency 1 and amplitude 9 to make little additional contribution to predicting the age criterion.

Recovery Functions

The adjusted R_2 values plotted in Figure 53 reflect amplitude and latency recovery. They were obtained after adjustment for covariance within groups for R_1, i.e., they are independent of the accompanying R_1 values. The R_2 value for each subject used in Figure 53 represents the mean of adjusted R_2's across the 19 interstimulus intervals. Analysis of variance indicated that mean R_2 amplitudes 1, 4, 6, and 9 varied significantly with age, as did mean R_2 latencies 1, 4, 5, 6, and 7. In older subjects, R_2 amplitudes 1 and 9 were more negative and R_2 amplitudes 4 and 6 were more positive. The significant R_2 latency relationships all reflected an increase with age.

For Figures 55 and 56, the subjects were divided into two groups, above and below age 40; mean adjusted R_2 amplitude and latency for each interstimulus interval are plotted for each age group. The analysis of variance showed significant interactions between age and interstimulus interval for R_2 amplitudes 4, 5, 6, 7, 8, 9, and 10 and for R_2 latencies 5, 6, and 7. The course of recovery thus varied significantly with age for 10 of the 18 variables. The rather different shapes of mean amplitude recovery curves for older and younger subjects can be seen clearly in Figure 55. Some of the results shown in Figure 55 may be related to the tracings of Figure 19. The negative dips in R_2 peaks 5, 7, and 9 in the records of Figure 19, taken from a 75-year-old man, appear to reflect the general trend indicated by the mean R_2 curves of older subjects for amplitudes 5, 7, and 9. It is possible that the results for R_2 amplitude 8 may reflect the increased negativity in peak 7, since the marked negative shift in peak 7 may not have been completed when peak 8 occurred; this would tend to move peak 8 toward the negative side.

A multiple linear regression analysis also was performed for the adjusted R_2 amplitude and latency values with age as criterion. Three values were employed for each amplitude or latency recovery curve. These were the means of the adjusted R_2 values obtained with interstimulus intervals ranging from 2.5 to 20, 25 to 60, and 70 to 120 msec. The initial multiple correlation coefficient between all 48 R_2 variables and age was 0.909. The final coefficient between the age criterion and eight residual variables was 0.810 ($p < 0.001$). These residual variables were; mean R_2 amplitudes of peaks 4, 5, and 6 for the 70 to 120 msec interstimulus interval range; mean R_2 latencies 1 and 5 for the 2.5 to 20 msec intervals; mean R_2 latencies 6 and 7 for the 25 to 60 msec intervals; and mean R_2 latency 8 for the 70 to 120 msec intervals. The results indicate that these eight R_2 measures could be combined to reduce the error of predicting age by about 65 %.

An additional regression analysis with age as criterion was carried out, beginning with the two R_1 and eight R_2 variables that remained after the separate R_1 and R_2 regression analyses. This analysis yielded an initial coefficient of 0.835 and a final one of 0.813 after four R_2 variables including all three amplitudes and latency 8 had been eliminated. This result suggests that the addition of R_2 variables would increase the accuracy of predicting

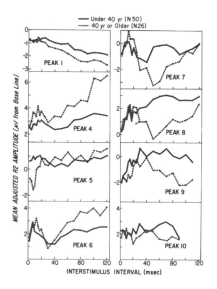

Fig. 55. Mean somatosensory amplitude-recovery curves for 76 nonpatient subjects divided into two age groups. Peak designation as in Fig. 24.

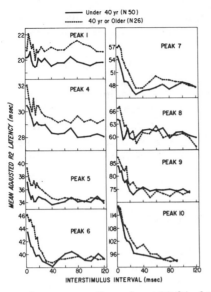

Fig. 56. Mean somatosensory latency recovery curves for 76 nonpatient subjects divided into two age groups. Peak designation as in Fig. 24. Note slower latency recovery in older subjects.

age by no more than about 10 % of that attainable with the two R_1 variables; conversely these two R_1 variables would add little to prediction of age from the eight R_2 variables.

Considering the possible sources of error in the detection and measurement of the various peaks in the evoked response, the high correlations obtained with age seem noteworthy. One may, however, question whether the biological effects suggested by them are central or peripheral. Latency differences with age may, in part, be due to peripheral nerve conduction velocity decrease with age (Norris *et al.*, 1953). However, since the R_2 values were adjusted for variations in R_1, the fact that there were more age differences in latency recovery than in R_1 latencies suggests that additional mechanisms must be operative. The R_1 amplitude variations with age could result from physical differences affecting conducting properties of structures between the recording leads and the underlying brain, i.e., scalp, skull, and meninges. Here also, however, the fact that recovery varies with age cannot readily be explained in terms of extracerebral differences or changes, since R_1 and R_2 should be equally affected.

SOMATOSENSORY RESPONSES AND SEX

Figure 54b shows composite mean somatosensory responses for nonpatient males and females. The age distribution in the two sex groups was the same. The only sex differences found significant by analysis of variance were the shorter mean latencies for peaks 1 and 6 in females. These latency differences are probably attributable to the shorter average length of the conduction pathway in women.

Analysis of the R_2 values in relation to sex revealed significant differences only for amplitude 4 and latency 1. Females showed less recovery of amplitude 4 and greater recovery of latency 1.

A multiple linear-regression analysis of R_1 amplitude and latency values against sex yielded an initial coefficient of 0.665 and a final one of 0.634 ($p < 0.001$). The residual R_1 variables were amplitudes 5, 8, and 9 and latencies 1, 4, 6, and 7. A multiple regression analysis employing the 48 R_2 variables yielded a nonsignificant initial R. The results indicate that, although sex must be controlled in studies of individual differences in somatosensory responses, variations with sex are probably not as great as those with age.

AUDITORY RESPONSES, AGE, AND SEX

Although visual responses are easy to record without averaging in the newborn, Ellingson (1964) reported consistent failure to record responses to clicks from electrodes placed over the sylvian region in over 100 neonates. The sylvian lead placement appears to be unfavorable, since even with an averaged-response computer, he recorded possible responses in only two of six newborns. In contrast, Barnet and Goodwin (1965) were successful in recording averaged responses to clicks in most newborn infants. The response was maximal at the vertex, and its most prominent component was a positive peak with an average latency of 267 msec. The amplitude of this large, late component varied linearly with stimulus intensity, while the amplitude and latency of the earlier components did not. As has been observed with somatosensory responses, the click evoked response of the newborn was of longer latency than that of adults. Barnet and Goodwin also compared responses during different stages of sleep and found that deep sleep was associated with larger responses of longer latency than was lighter sleep. Figure 57 shows the relationship between the intensity of the click and the amplitude of the large late component for 12 of the newborn infants in Barnet and Goodwin's study.

Auditory evoked responses have also been studied in prematurely born infants by Weitzman et al. (1967), Schulte et al. (1967), and Akiyama et al. (1969). Weitzman et al. noted that the amplitude and latency of the components differed according to the location of the recording electrodes and the maturational and behavioral state of the infant. In infants of 25 to 28 weeks estimated gestational age, there was a diffuse scalp response with a negative wave at 200 to 270 msec followed by a large positive wave at 700 to 900 msec latency. As maturation progressed, the latencies of all wave components decreased and by 34 to 37 weeks gestational age, the pattern approached that seen in full-term infants. A positive component of 90 msec latency appeared and a very prominent wave with a latency of about 300 msec was present at the midline electrodes. Differences between

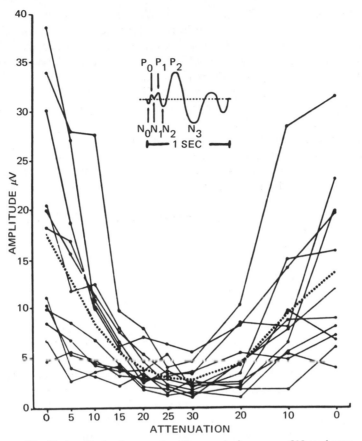

Fig. 57. Amplitude of averaged auditory evoked response of 12 newborns as a function of click intensity. Amplitude measures refer to P_2-N_3 in insert of a typical response. Dotted line is mean for all subjects. Upward deflection indicates positivity at vertex with respect to T_5. (Reprinted with permission from Barnet and Goodwin, 1965.)

active and quiet sleep were recognizable by 33 to 36 weeks gestational age. Schulte *et al.* (1967) and Akiyama *et al.* (1969) also found differences between responses in active and quiet sleep in prematures with gestational age from 32 to 37 weeks.

Data concerning postnatal development of auditory responses have been reported by Ferriss *et al.* (1967a) who made longitudinal observations in 20 normal full-term infants from birth to the age of 6 months. They found that the initial long latency vertex wave increases in prominence during the first two months. Barnet and Lodge (1967) reported amplitude measurements for $P_2 - N_3$ (Figure 57) and latency measurements for P_2 in 15 normal infants from the newborn period to 14 months of age. There was a definite trend toward increase in amplitude and decrease in latency as the children developed. Ornitz *et al.* (1967) compared auditory evoked responses during two sleep stages in both children and adults. They found that the younger children showed a simpler evoked-response wave form than the adults. Suzuki and Taguchi (1968) considered the wave shape of auditory responses during sleep to be similar in children and adults. These authors noted that, for a given stage of sleep, amplitudes were greater and latencies less in adults compared with children aged 16 days to 3 months, while values were intermediate in children aged 1 to 5 years. The latency figures given by Rapin (1964) for adults and children suggest that the later components have a somewhat longer latency in children than in adults. Price and Goldstein (1966) also observed latency decrease with increasing age in a group of children, but their data suggested that amplitude decreases with age.

Sex differences in auditory response characteristics do not appear to have been reported.

VISUAL RESPONSES AND AGE

Infancy and Childhood

Ellingson (1964) found that reasonably consistent and identifiable evoked responses to single flashes of light were recordable in the unaveraged EEG of about 50 % of neonates. The best lead for recording these responses was one placed just above the inion. Ellingson reported the amplitude of the occipital response to be generally higher in the newborn than in older children and adults, but pointed out that this may be due to the less prominent background activity in the infants. He found the evoked potentials in newborns to be characterized by "fatiguability," noting that the interval between stimuli required to avoid response reduction was between 1 and 3 sec. Ellingson's mean value for latency, measured to the peak of the first positive phase in the infant response, was 190 msec or about twice that of an adult. Figure 58 shows the relationship between this latency measurement and age. It will be noted that the latency is longer in premature infants. An interesting aspect of the curve is the break occurring between the fourth and fifth weeks post-term. Ellingson (1960) found a similar curve in relation to body weight, with the break coming between 8 and 9 lb. He suggested that the change in character of the curve may be associated with a growth spurt in the visual system, or that it reflects the development of two systems, the scotopic and photopic. The latter possibility is discounted by recent data of Fichsel (1969) who compared responses to white and colored light in infants and children of various ages. He found that although latencies were prolonged with colored light, their variation with age paralleled that with white light.

It appears that the definitive form of the visual response is present from about the sixth year on (Weinmann *et al.*, 1965). The development

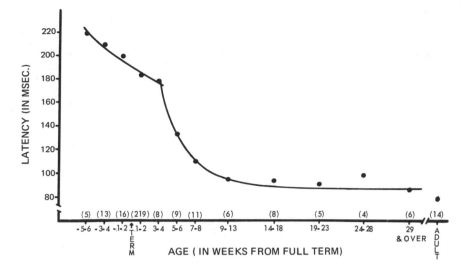

Fig. 58. Plot of mean visual evoked-response latencies against age. The numbers in parentheses just above the abscissa represent the numbers of infants upon whom the dots above are based. The dot at far right is the mean latency of adult controls. (Reprinted with permission from Ellingson, 1960.)

of the response in infants and children is depicted in Figure 59, which shows superimposed tracings obtained by Creutzfeldt and Kuhnt (1967) in various age groups. They classified two different response patterns in all age groups (Columns A and B) and an additional type (Figure 59C) in the age groups from 5 to 14 years. Figure 59D shows mean potentials from small samples of different ages. Clearly, the older the child, the earlier the peaks. Amplitude of at least the late component appears to be maximal in the age range from 3 to 9 months. However, this observation concerning amplitude is not confirmed by the quantitative studies of Dustman and Beck (1966). These authors measured the total amplitude excursion during the first 250 msec of the visual response; they also made measurements for each half of this time period. Their 215 subjects ranged in age from 1 month to 81 years. Figure 60 shows their results. The responses recorded from the occiput increased rapidly in amplitude reaching a maximum in the 5- to 6-year-old group. The amplitude then declined from age 7 until age 13 when there was another abrupt increase in amplitude.

Adolescence and Adult Years

The data of Dustman and Beck indicate that visual-response amplitude appears to stabilize at about age 16, and to remain stable till about age 50. Although total amplitude did not change significantly in older subjects, significant reverse trends were noted in the separate data for earlier (0–

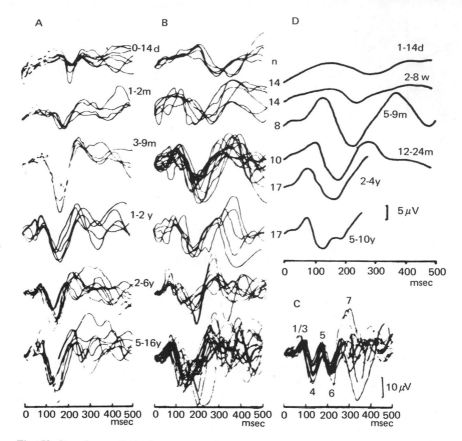

Fig. 59. Superimposed visual responses of subjects in different age groups (d = days, w = weeks, m = months, y = years). The potentials are divided into different groups *(A–C)* according to their shape, but this classification is not systematic and does not necessarily hold for every individual throughout the different ages. *D:* mean potentials at different ages taken from small samples (n on the left of each mean potential is the number of individuals. (Reprinted with permission from Creutzfeldt and Kuhnt, 1967.)

125 msec) and later portions (126–250 msec) of the response (Figure 60). Amplitude of the 0–125 msec portion was greater in older than in younger subjects, but the reverse was true for the later portions (126–250 msec). The increase of amplitude with age in the earlier portion is consistent with the results of Kooi and Bagchi (1964*a*) who found low, but significant, positive correlations between age and certain amplitude and latency measurements in an adult sample. Their findings suggest that both amplitude and latency tend to increase with age.

Our own studies of visual responses gave somewhat equivocal results with respect to age. The R_1 values obtained in a study of visual recovery functions in a mixed group of adult psychiatric patients and nonpatients

(Shagass and Schwartz, 1965*a*) yielded significant age differences for the amplitude measured between peaks 3 and 4 (Figure 41 and 42). The lowest amplitudes were in the 20- to 29-year-old group; the mean values by decade suggested a *V*-shaped curve. The degree of after-rhythm did not vary with age. No age differences were, however, found in data obtained in the same subject group, which consisted of single averaged responses to three different flash intensities (Shagass *et al.,* 1965*a*). The significant results obtained in the recovery study may reflect the fact that the values used were the means of 23 R_1 measurements per subject and, therefore, more reliable.

In another study (Straumanis *et al.,* 1965), responses to flash of 18 healthy subjects with a mean age of 72 years were compared with those of 18 nonpatient subjects (mean age 24 years) studied by Shagass and Schwartz (1965*a*). The composite mean evoked response obtained by measuring eight sequential latencies and amplitudes in these two age groups are shown in Figure 61. Statistical analysis indicated that the older subjects had a significantly longer latency for all of the first six peaks and significantly greater amplitude, measured as deviation from base line, for peaks 1 and 2. The amplitude of peak 5 was also significantly more negative in the older group. The increased amplitude and prolonged latency of early peaks in the older subjects agrees with the results of Dustman and Beck (1965*a*; 1969). Floris *et al.* (1969) have also found increased amplitude and latency in visual responses of older adult subjects.

We found no significant relationships between age and measures of visual response recovery functions (Shagass and Schwartz, 1965*a*), but Floris *et al.* (1968; 1969) report increased facilitation in older subjects, thus providing results in substantial agreement with those obtained by us for somatosensory recovery.

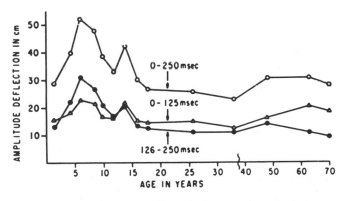

Fig. 60. Relationship of visually evoked-response amplitude to age. The measure of amplitude is the excursion in centimeters of wave components occurring during the first 250 msec, first 125 msec, and 126–250 msec of responses recorded from occipital scalp. (Reprinted with permission from Dustman and Beck, 1966.)

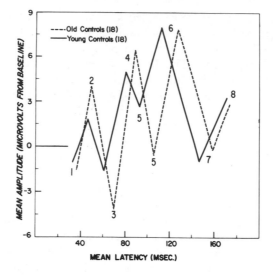

Fig. 61. Composite mean visual evoked responses in old and young nonpatient subjects. Note higher amplitudes and more prolonged latencies in older subjects. (Reprinted with permission from Straumanis *et al.*, 1965.)

VISUAL RESPONSES AND SEX

Sex differences in visual responses were reported by Rodin *et al.* (1965). They found higher amplitudes and shorter latencies in females than in males. Our results were generally similar to those of Rodin *et al.* (Shagass and Schwartz, 1965a). The amplitude measured between peaks 3 and 4 (Fig. 41) was greater in females, and the latencies of peaks 1, 2, and 3 were significantly shorter. Perhaps the most interesting sex differences were found in relation to the recovery functions. The statistical analysis was based on R_2 values adjusted for their covariance with R_1. All of the three adjusted R_2 amplitude measurements were significantly greater in the females at $p < 0.01$. The amplitude-recovery curves of the female subjects were all steeper than those of the males. The recovery results suggest that women have greater cortical "excitability" than men. It may be noted that this finding is in the opposite direction from that obtained with somato-sensory recovery functions; in the latter, female subjects showed less amplitude recovery. The greater recovery of visual responses in females may be related to the observation that women show higher-amplitude photic driving (Shagass, 1955a).

CONTINGENT NEGATIVE VARIATION

Walter (1969) summarized CNV findings in 60 normal children. Although CNV waves were observed in very young children aged 3 years or less, these were transitory. In most children till the age of 7 to 8 years, large secondary waves were observed, but these did not seem dependent upon stimulus association or voluntary participation. From about age 7, typical CNV responses began to appear, but the fully developed CNV was not seen in all normal subjects until about age 20. Walter notes that, as the CNV matures, it becomes less susceptible to social influences; he has observed that "accidental" factors, such as the color of the experimenter's clothes, can affect CNV size in children. Lairy and Guibal (1969) compared CNV in children with normal and defective vision, ranging in age from 11 to 18 years. They noted that younger children did not have CNV's under all stimulus conditions to the same extent as older children. These variations would be in accord with Walter's observations.

Low et al. (1966) studied CNV in subjects as young as 4 years of age. They found rather large CNV potential shifts in the three children under twelve years old, but after age 12 noted no correlation between CNV amplitude and age. Another characteristic observed by Low et al. in the CNV of young children is that the negative shift persisted for a considerable period of time after the response to the imperative stimulus. One of their illustrations shows the response persisting for several seconds. The possibility that such prolongation of the negative shift may be characteristic of younger children is of interest in relation to the results of Timsit et al. (1969), which suggest that persistent negativity following button press may be a response independent of the CNV. They have called this response the C wave.

Sex differences in CNV characteristics do not appear to have been reported.

HANDEDNESS

Eason and White (1967) found that evoked responses recorded from the right occipital lobe, due to stimulation of the nasal field of the right eye, were consistently smaller than those obtained from temporal stimulation of that eye. Subsequent to this observation, Eason et al. (1967a) investigated the responses to retinal sites symmetrically located in opposite hemispheres. Of the three subjects, two were left-handed, and in these they found larger amplitude responses from the right occipital lobe, when it was the primary response area than from the left occipital lobe, when it was primary. Eason et al. studied 26 additional subjects, of whom 13 were right-handed and 13 were left handed. The visual responses obtained in these subjects are shown in Figure 62. It may be seen that, for most of the left-handed subjects, the response of the right lobe was of greater amplitude than the response of the left lobe. This relation did not seem to be present in the right-handed subjects.

These observations suggest that it may be relevant to determine the handedness of subjects in studies of visual responses, since the presence of left-handedness seems to present special problems.

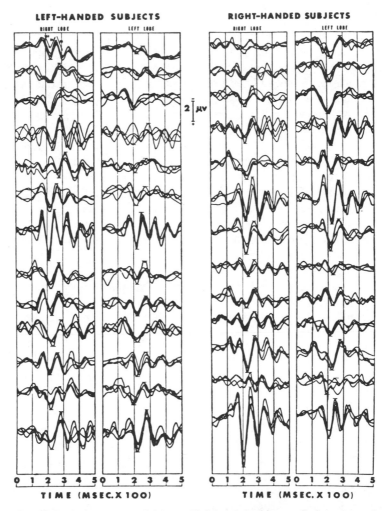

Fig. 62. Evoked responses of right- and left-handed subjects to flashes presented in right- and left-visual fields. Note asymmetry in left-handed subjects with larger responses in right lobe than in left lobe. (Reprinted with permission from Eason *et al.*, 1967*a*.)

TIME OF DAY

Heninger *et al.* (1969) made observations on diurnal variations in the characteristics of somatosensory, auditory, and visual evoked responses as part of a study of possible relationships between 17-hydroxycorticos-teroids and CNS activity. Recordings were made from six normal male subjects in four identical two-hour recording sessions which began at 7 AM and 4 PM on two separate days at least one week apart. Half of the subjects received dexamethasone for the first session and a placebo for the second, while the order of drug administration was reversed in the remaining subjects. The results showed that somatosensory peaks designated as 6 and 7 by Heninger *et al.*, which correspond to peaks 8 and 9 in our scheme, both increased significantly in latency from morning to evening. Conversely, peaks 9 and 10, of Heninger *et al.*, which have average latencies of about 150 and 225 msec, both showed reduced latency in the evening. Somatosensory amplitude 7 to 8 in the Heninger *et al.* scheme decreased significantly. No significant changes were observed in the auditory response. The visual response recorded from the vertex showed a decreased latency in the evening for a peak occurring at about 200 msec and the occipital visual response manifested an evening decrease in three of the four amplitudes measured. The administration of dexamethasone, which suppresses release of ACTH and imposes a stable state with respect to 17-hydroxyketosteroids, did not significantly affect the observed diurnal changes in the evoked responses. Heninger *et al.* concluded that the diurnal variations were not a function of diurnal changes in either ACTH release or circulating 17-hydroxycorticosteroid levels.

These observations of diurnal changes in several evoked-response characteristics point to the need to control for the time of day during which recordings are made. This seems to be particularly necessary when the evoked-response data of interest are to be measured in components other than the very earliest.

RESPIRATORY AND CARDIAC CYCLES

By triggering flash stimuli at specified points in the respiratory and cardiac cycles, Callaway and Buchsbaum (1965) demonstrated visual evoked response differences related to the point in the cycle. Responses evoked during inspiration were more similar to one another than they were to responses evoked during expiration. Responses evoked during the same phase of the cardiac cycle were more similar than responses evoked at different phases. These findings were interpreted as indicating that cardiac and respiratory cycles can contribute to evoked response variability. Since the effects were not large, and since the influence of these cycles is probably fairly well randomized in the usual recording situation, the effect of these factors is probably minimal under ordinary circumstances. However, if the stimulus should for some reason become significantly time-locked to a particular phase of the respiratory or cardiac cycles, important effects can occur. This is not only because of the genuine evoked-response variations associated with these biological cycles, but because the electrical events associated with them may be averaged together with the EEG. For example, we have found that triggering stimuli from the R wave of the EKG led to significant averaging of the R and S waves. These are apparently regularly recordable from the scalp but they are brought out only by averaging. Callaway and Buchsbaum (1965) also noticed this and found it necessary to subtract nonstimulus sweeps, triggered at the same point in the cycle as the flash, in order to cancel the effect.

COMMENT

The methodological implications of the material presented in this chapter seem clear. Investigators must control for age and sex in studies attempting to relate evoked-response characteristics to other criteria. They also need to pay more attention to handedness and time of day as sources of variance in results and to avoid locking stimuli to particular phases of biological cycles.

Latency appears to be the evoked-response characteristic that is best correlated with age. Latencies are longest during early life, shorten progressively as development proceeds, are minimal during young adulthood, and lengthen again during old age. The developmental processes that may underlie the latency changes of early life include myelinisation, thickening of fibers, shortening of the transmission time of synapses, and an increase in the number of dendrites and their connections in the CNS. The prolongation of latencies in later life may reflect decreased nerve conduction velocity and the loss of central neurones (Brody, 1955); the latter process would be accompanied by a reduction in the number of CNS connections.

Although peripheral nerve change may be a determinant of latency prolongation in later life, it seems likely that central mechanisms related to aging are also involved in both the latency and amplitude variations with age. One relevant supporting datum is the finding that somatosensory latency recovery varies more with age than does latency of the first response, even though the recovery measure is independent of R_1 variations. Another is the fact that peripheral nerve recovery is poorly correlated with cortical recovery (Shagass and Schwartz, 1964a). Also, latency and amplitude measures, including those for recovery functions, both increase with age. Since amplitude increase with variations of stimulus strength in an individual subject is accompanied by latency *decrease* (Diamond, 1964), the relations with age seem paradoxical. If there were no additional central factor, one would expect latency prolongation in old age to be associated with *reduced* amplitude. Such an expectation would also be in accord with the idea that longer latencies allow greater possibility for dispersion of responses in time, so that averaged amplitudes would be smaller.

Amplitude can be interpreted as the resultant of the combined excitatory and inhibitory activities in the neuronal aggregate giving rise to the response. The fact that amplitudes increase rather than decrease with age suggests that there is an alteration of the balance between excitatory and inhibitory CNS activities as aging progresses. Increased amplitude could reflect either more excitation or less inhibition. The hypothesis that inhibitory activities may be decreased in old age is favored by clinical evidence of greater sensitivity to CNS depressant drugs in old people (Goldfarb, 1967). Also, if one accepts Grundfest's (1959) concept that inhibitory mechanisms underlie the precision of nervous activity, the poorer perceptual functioning of old age should involve reduced inhibitory activity. Thus, although by no means proved, the idea that amplitude increases with age as a consequence of reduced central inhibitory activity seems tenable. One may further speculate that the neurons lost with aging include more units subserving inhibitory than subserving excitatory functions.

Sex differences in visual responses follow the expected relations between amplitude and latency; females have both greater amplitudes and shorter latencies than males. Sex differences in length of the visual conduction pathway seem insufficient to account for the latency findings. Since the investigations in which the sex differences were found (Rodin *et al.,* 1965; Shagass and Schwartz, 1965a) did not control pupil diameter, it could be that female pupil size was greater, resulting in more intense retinal stimulation. However, since Kooi and Bagchi (1964a) found no correlation between visual-response latency and pupil size, it seems possible that the results may reflect a sex difference in retinal or cerebral functioning. More carefully controlled studies are needed to specify the mechanism of the sex differences in visual response latency.

We found that amplitude recovery of visual responses was greater in females (Shagass and Schwartz, 1965a). This result seems unlikely to be due to peripheral factors, since the R_2 values were adjusted for differences in R_1. Sex differences in visual recovery were opposite in direction to those for somatosensory recovery, which was less in females. Thus, although women showed greater cortical "excitability" in the visual sphere, they manifested less in the somatic modality. The functional correlates of these sex differences in recovery remain to be established. The modality differences suggest that it may be worthwhile to compare the sexes in relation to perceptual abilities in various sensory spheres and to correlate these behavioral measurements with recovery function determinations. The variation with sensory modality in correlations with sex also fails to support the possibility that sex differences in temperament and personality may involve general differences in "cerebral excitability."

Chapter 5

Evoked Responses and Impaired Consciousness

The EEG is generally accepted as a good indicator of gross level of alertness. Alerting stimuli desynchronize the alpha rhythm recorded during the resting awake state, and sleep is associated with marked EEG changes. When evoked-response recording became possible, investigators soon sought to determine whether the responses also reflected level of alertness. This chapter will deal with states of impaired consciousness: sleep, delirium, and coma. Phenomena related to states of heightened alertness will be considered in the subsequent chapter.

SLEEP STAGES

The dramatic EEG changes associated with sleep were observed from the earliest days of electroencephalography (Loomis *et al.*, 1935; 1938). Sleep research was intensified by the discovery of the phase associated with rapid eye movements (REM) and dreaming (Aserinsky and Kleitman, 1955; Dement and Kleitman, 1957). Subjects awakened during the REM phase of sleep were usually able to report that they had been dreaming just prior to awakening, whereas, such a report was rarely obtained when they were awakened during other phases. The REM phase is characterized by a low voltage, fast EEG, the occurrence of rapid eye movements as seen in EOG recordings, reduction of activity in the EMG signifying reduced muscle tone, and a number of other physiological changes (Dement, 1967). The EEG, EOG, and EMG have usually been employed to provide criteria for classifying stages of sleep in evoked-response studies. The most commonly used classification is that of Dement and Kleitman (1957). This classification contains five stages, I to IV and REM. Stage I involves the same EEG patterns as REM, but no eye movements or EMG reduction. The stage II EEG is characterized by spindle-shaped trains of 12 to 18 c/sec waves. In stage III, up to half of the EEG record is occupied by slow waves of 3 or fewer c/sec. In stage IV, more than half of the EEG consists of high-amplitude slow waves.

SOMATOSENSORY EVOKED RESPONSES AND SLEEP

Uttal and Cook (1964) and Giblin (1964) reported changes in the responses to peripheral nerve stimulation during sleep in a few subjects. Giblin showed a generalized reduction in amplitude of the response. Uttal and Cook found a tendency to slowing of the late component on the hemisphere contralateral to the stimulus and virtual abolition of the slow component recorded ipsilaterally. Nakagawa (1965) found little change in the amplitude of the initial component during sleep, while four later components, with latencies from 50 to 200 msec, changed markedly in amplitude; no latency data were reported.

More systematic studies were carried out by Goff et al. (1966), who recorded from 11 subjects for a total of 27 nights. They found that the latency of the initial negative and the first two positive peaks increased progressively from waking and REM through stages II, III, and IV of the Dement and Kleitman classification. The latency increases are reported as statistically significant only for component 1, which occurred about 10 % later in stage IV than during waking. During stages III and IV, the amplitude of the first two positive components was significantly increased over that measured during REM, and the initial positive component was smaller during REM than during waking. Goff et al. also examined the myogenic response recorded over the trapezius muscle in three subjects and found no change with sleep.

Goff et al. noted that the most dramatic evoked-response changes with sleep were in the long latency components recorded from an electrode placed at the vertex. Components 4 and 5 in their scheme (Figure 23) were markedly reduced during early drowsiness and appeared to be absent during REM sleep. However, commencing in stage II and becoming most prominent in stages III and IV, very late potentials appeared in the form of two negative–positive sequences. The latest of these occurred 700 msec or more after the stimulus and the latency was longer and the amplitude larger in the deeper stages of sleep. These long latency components of the somatosensory sleep responses were similar to those evoked by sound in the same subjects. Furthermore, they appeared to resemble waves observed by Kooi et al. (1964) in response to flash. In common with other workers, Goff et al. interpreted the large, long latency responses during sleep, which do not appear to be sensory modality specific, as corresponding to the K-complex described by Loomis et al. (1938).

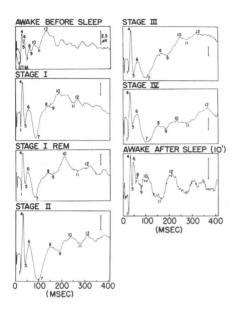

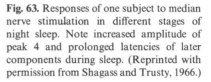

Fig. 63. Responses of one subject to median nerve stimulation in different stages of night sleep. Note increased amplitude of peak 4 and prolonged latencies of later components during sleep. (Reprinted with permission from Shagass and Trusty, 1966.)

The somatosensory response data during sleep obtained in our laboratory agree in most respects with those of other workers, but differ in some details (Shagass and Trusty, 1966). Recordings in all stages of sleep, including stage I without REM, and waking measurements made before and after sleep, were available for seven subjects. Figure 63 shows tracings from one of these. For descriptive purposes, the various peaks were numbered sequentially. Although the initial deflections are almost certainly identical during sleep and waking, the identity of the later peaks with the same number is open to question. The amplitude and latency measurements were subjected to analysis of variance. All ten latency measures showed significant increases with sleep ($p < 0.001$). Significant amplitude changes were found for peaks 1, 4, 9, 10, 11, and 12. During sleep, the initial two peaks increased in amplitude, peaks 9 and 10 showed increased relative negativity, and peaks 11 and 12 demonstrated increased positivity. Because our analysis time was not sufficiently long, we did not observe the very late waves reported by Goff *et al.* (1966). However, the changes during stages III and IV for peaks numbered from 8 onwards probably reflect some of the late events noted by these workers.

Comparisons by *t*-tests between individual sleep stages were made for variables found significant in the analysis of variance. As might be expected, the largest number of differences were found between waking and stage IV However, with one exception, responses recorded during contiguous sleep stages also differed significantly with respect to most of the latency measurements. The exception was provided by the virtual absence of difference between the REM phase and stage I. The responses evoked during REM differed from those of waking and stage II, but resembled the latter more than those of waking. We made additional analyses utilizing all available data from the 21 subjects who were studied, i.e., including 14 subjects for whom recordings were missing for one or more stages. Comparing the measurements during different stages of sleep, the significant differences found for the sample of seven subjects were confirmed and augmented in number. However, although many variables discriminated between adjacent sleep stages, almost none distinguished between stage I and the REM phase.

It was of particular interest that a number of significant differences, particularly in latency, were found between the waking state recordings obtained before and after sleep. Latencies of peaks 5 to 10 were significantly longer in the waking record after sleep than before. The tracings of Figure 64 illustrate this. They were made during daytime. In deep sleep, the response was remarkably altered from that recorded before the subject went to sleep. The subject was awakened and about 5 min later the response shown in Figure 64C was obtained. Although of similar general pattern to that obtained during presleep waking, the latencies of the later components were still considerably prolonged. Even after 20 min of waking, the latency of peak 6 had not returned to the waking value (Figure 64D).

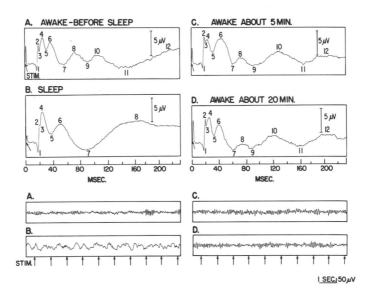

Fig. 64. Somatosensory responses of one subject to median nerve stimulation during daytime sleep. EEG samples taken during correspondingly lettered averaging sequence are shown below. Note particularly the delayed return of presleep latencies of peaks from 6 on after subject was awakened. Subject reported feeling "groggy." (Reprinted with permission from Shagass and Trusty, 1966).

Saier *et al.* (1968) also studied somatosensory responses during sleep. Their technique differed from ours in that they used a "monopolar" montage and their averager was apparently unable to provide adequate detection of the very earliest response events. However, their results are in general agreement with those that we obtained. They found that the latency of their earliest positive deflection was markedly prolonged during sleep and that its amplitude was increased, particularly during the later stages. In ten subjects, they also noted that the large positive wave occurring between 200 and 250 msec during waking was inverted in phase and became negative during sleep associated with slow waves in the EEG. They found that responses recorded during the REM phase were more similar to those obtained during waking than during other stages of sleep.

Saier *et al.* also drew attention to the state of "grogginess" or "cloudiness" (*obnubilation*) found upon wakening. In agreement with our results (Figure 64), they found that the slowing of later activities persisted for some time after wakening. They noted that, although the postsleep recovery of the evoked-response characteristics was gradual in recordings from the "specific regions," the responses recorded from the vertex assumed the waking pattern almost immediately after the subject was roused.

The main findings obtained in studies of somatosensory response during sleep may be summarized as follows: (1) in progression from waking to stage IV of sleep, latencies become increasingly prolonged; (2) the responses occurring during REM sleep are more like those of lighter than deeper stages of sleep; (3) during sleep from stage II on, there are late slow components which may be identical with the *K*-complex; (4) after awakening, the return of evoked response characteristics to the levels observed before sleep tends to be gradual; (5) amplitude changes are not as consistent as those for latency but, in general, the amplitude of the early components tends to increase during slow wave sleep.

The fact that the evoked response returns only gradually to its presleep state after awakening seems to be important for studies of individual differences. Apparently, the evoked response may vary considerably as a function of relatively minor variations in state of alertness. Some of the variations with time of day noted by Heninger *et al.* (1969) may have arisen on the basis of diurnal variations in alertness. Unfortunately, experimental control of alertness involves unsolved technical problems. There can be no assurance that experimentally imposed conditions of alertness produce the same effect in different subjects or in the same subject at different times. It would be desirable to have an independent criterion of alertness to monitor throughout the course of evoked-response recording. Although we had hopes that the quantified EEG might provide such a criterion, recent data suggest that these hopes may not be fulfilled (see Chapter 7).

AUDITORY RESPONSES AND SLEEP

Vanzulli *et al.* (1961) found that the latency and amplitude of click-evoked responses increased during sleep and that the changes were greatest when the sleep EEG was predominantly composed of slow waves. The findings reported by Williams *et al.* (1962, 1964) are more complex. Using a vertex-left occipital bipolar derivation, they found that the complex first component merged into one large wave and increased in amplitude during sleep, while the positive component at 170 msec decreased in amplitude. Between stages II and IV, later components became more predominant. During REM the early components were of about the same amplitude as found at sleep onset, whereas average peak-to-peak amplitudes were smaller during REM than in waking, or in any other stage of sleep. Williams *et al.* (1964) presented data suggesting that the auditory responses during stage I REM were similar to records taken while the subject was reading. During sleep stages II to IV, Nakagawa (1965) found increased positivity of auditory components occurring at 100 and 170 msec in the waking state. The values during REM sleep resembled those of waking.

Weitzman and Kremen (1965) made systematic measurements of amplitude and latency of click-evoked responses during sleep. Some typical records obtained by them are shown in Figure 65. They found that the

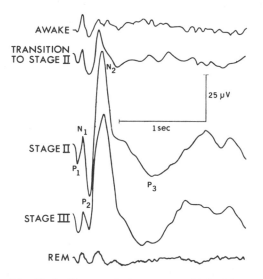

Fig. 65. Auditory evoked responses obtained in a subject awake and in various stages of sleep. One hundred stimuli, one every four seconds. Vertex-occipital recording. Note marked increase of amplitude in sleep except for stage REM. (Reprinted with permission from Weitzman and Kremen, 1965.)

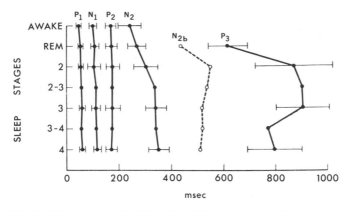

Fig. 66. Mean and standard deviation of the peak latency of each component of the auditory evoked response, awake and during various stages of sleep. Components as in Fig. 65. Latency of N_2 is significantly greater in all stages of sleep as compared with the awake and REM stage. (Reprinted with permission from Weitzman and Kremen, 1965.)

amplitude of the later components, N_2 and P_3, increased progressively from the awake state through the four stages of sleep. During the REM phase, amplitude was low and similar to that of the awake state, while the highest amplitude occurred during stage IV. The mean latency values obtained by Weitzman and Kremen are graphed in Figure 66. Although the general trend was for latency increase from awake to stage IV, these were statistically significant only for components N_2 and P_3 and only when comparing REM with other sleep stages.

Weitzman and Kremen also presented evidence indicating that the large late component found during sleep is probably identical with the K-complex. Although the K-complex has been described as most pronounced in the EEG during stage II sleep, they noted that it was clearly brought out by averaging during stage IV. Apparently the K-complex is present during stage IV, but obscured by the high-voltage slow activity.

The findings of Suzuki and Taguchi (1968) in children aged 16 days to 5 years are essentially similar to those of Weitzman and Kremen. They showed increasing amplitude and latency and the appearance of the late positive wave with increasing depth of sleep.

As with somatosensory responses, it appears that the main changes occurring in auditory responses with sleep are: (1) prolongation of latency; (2) appearance of long latency slow potentials, probably representing the K-complex; (3) general tendency to increase in amplitude, but with some variability; and (4) responses during the REM stage that are more similar to those of waking than to those found in deep sleep.

VISUAL RESPONSES AND SLEEP

Vanzulli *et al.* (1960) found that visual response latencies were increased and positive waves augmented in sleep; they were not able to record the earliest components adequately. Cigánek (1961*b*) noted increased latency of the earlier components and an increase in amplitude of his wave V, as well as delay in this component. He observed *K*-complexes during sleep and found that the after-rhythms were abolished. Brazier (1960) and Barlow (1960) also noted abolition of the after-rhythm.

Kooi *et al.* (1964*b*) conducted a very careful study of flash-evoked responses in which the subject's eyes were maximally dilated and, in addition to EEG evidence of sleep stage, level of awareness was evaluated by requiring the subject to press a key. They made the interesting observation that variation in the intensity of the stimulating flash resulted in much clearer differences in amplitude of the response when the subject was asleep than when he was awake. While the brightest flash produced the largest response during sleep, amplitude during waking was about the same for their least and most intense flash. During sleep, Kooi *et al.* found relatively little change in the latencies of earlier components, but those of later components were prolonged. The negative component at the occiput, observed at 60–100 msec during waking, was typically reduced in amplitude and, in some subjects, a positive deflection developed within this latency range. A later negative event was generally increased in amplitude.

Figure 67 shows typical changes in visual evoked responses observed during sleep in our laboratory (Shagass and Trusty, 1966). The most striking alteration occurred in the afterrhythm, which was abolished as soon as sleep began and did not return until waking. The latencies of nearly all peaks increased progressively from waking to stage IV. This conclusion was substantiated by analyses of variance performed on the data of five subjects for whom recordings were available in all stages of sleep as well as in the waking state before and after sleep recording. Of the twelve latencies, all but one, that for peak 4, yielded a significant *F*-ratio. On the other hand, only four of the twelve amplitudes, 1, 6, 10, and 11, varied significantly across stages of sleep. Visual peak 1 increased in amplitude, whereas peak 6 decreased. The changes in peaks 10 and 11 reflect increased positivity, probably in association with the late wave related to the *K*-complex (Figure 67). Comparison of different waking and sleep stages by *t*-test showed that the REM phase was not differentiated from stage 1 without REM, whereas all other stages differed with respect to many variables, particularly measures of latency. An analysis using all available data from the total subject group to compare individual sleep and waking stages, confirmed and extended the findings obtained in these five subjects.

Comparison of the waking records taken before and after sleep in Figure 67 shows differences somewhat similar to those depicted for the somatosensory response in Figure 64. Five minutes after the subject was

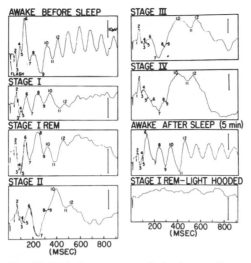

Fig. 67. Evoked responses to flash of one subject, awake and during various stages of sleep. Note disappearance of afterrhythm from stage I on and similarity of responses during stages I and I REM. (Reprinted with permission from Shagass and Trusty, 1966.)

awake, the later peaks appeared to be still quite delayed. Statistical comparisons of the two waking stages revealed that amplitudes 4 and 9 were smaller after sleep than before and that latencies 3, 4, 6, 7, 8, 9, 10, and 11 were significantly longer. However, as Figure 67 indicates, the afterrhythm returned shortly after the subject was roused.

Saier *et al.* (1968) found that a marked positivity took the place of all components existing in the waking state during the first 150 msec early in sleep and reached its maximum in stage IV. They found this in all derivations and in all of their 29 subjects. Saier *et al.* suggest that other authors may not have observed this effect because they did not use monopolar derivations. Since Kooi *et al.* (1964*b*), who also recorded monopolarly, did note a tendency toward replacement of negativity by positivity in the 60–100 msec latency range, this explanation seems to be supported. Saier *et al.* also noted a large negative wave at about 250 msec. A similar wave was observed in the infant by Kassabgui *et al.* (1966). The tracings for stages II, III, and IV, in Figure 67 also show a large negative wave at about 250 msec latency.

COMMENT

Evoked responses in all sensory modalities show marked changes during sleep. The changes appear to be most consistently depicted by latency measurements. Our own latency results for somatosensory and visual responses are shown in Figure 68; they generally resemble those for auditory responses shown in Figure 66. Latencies during all stages of sleep tend to be longer than those during waking and they tend to increase progressively from stage I through stages REM, II, III, and IV. As with amplitude findings, the latency differences between stages I and REM are slight and not significant. Although the absolute magnitude of latency changes in the earliest peaks was small, the fact that they were statistically significant is important because there is little reason to doubt that the initial deflections represent the same components in both waking and sleep. The later components showed much greater latency shifts, but it is impossible to eliminate doubts concerning the identity of these peaks between waking and sleep.

Amplitude findings appear to be much more variable from study to study. Some of the variation may be due to different lead placements. In our own results, it appeared that there were essentially three kinds of amplitude differences between waking and deep sleep: (1) the initial components (1–4) became larger in sleep; (2) the latest peaks measured (somatosensory 11 and 12; visual 9 to 12) showed increased positivity, probably because they were involved in positivity related to the K-complex; and (3) the intermediate peaks tended to decrease in amplitude or become more

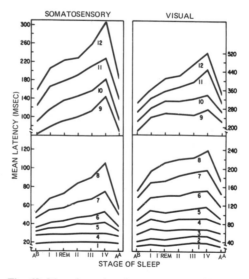

Fig. 68. Mean latencies of somatosensory ($N = 7$) and visual ($N = 5$) response peaks recorded during waking and various stages of sleep. Peaks numbered as in Figs. 63 and 67. A^B and A^A, awake before and after sleep, respectively. (Reprinted with permission from Shagass and Trusty, 1966.)

negative. Increase in amplitude of the earliest components is an important finding. It provides assurance that uncontrolled experimental variables, such as arm movement leading to altered contact of the stimulating electrodes, undetected head movement away from the light, pupillary constriction, or rotation of the eyeballs were not responsible for the evoked-response changes. All of these factors could reduce stimulus intensity and lead to prolongation of latency, but this should have led also to a reduction of response amplitude instead of the increase which was actually found. Increased amplitude also argues against increased latency variability during sleep, since such variability should decrease amplitude. Additionally, increased amplitude seems discordant with the idea that afferent input to the cortex may be reduced during sleep.

The REM phase of sleep has been considered to be the "deepest" stage of sleep by most workers studying animals such as the cat. The data from human evoked-response studies are, however, nearly unanimous in indicating that responses obtained during the REM phase are closer to waking than they are to those obtained during stage IV sleep. The discrepancy suggests species differences in organization of electrophysiological activity during sleep, but it may be noted that not all experimenters with animals regard the REM phase as the deepest sleep stage. For example, Winters (1964) considers REM to be a transitional zone between wakefulness and sleep.

The reduction in amplitude and the slowing of intermediate peaks, together with their delayed restitution after waking, suggests that these peaks may reflect activities related to alertness. The subject of Figure 64 stated that she felt "groggy" after being awakened. Saier et al. (1968) referred to a "cloudy" stage. The latencies of these peaks may possibly provide sensitive indicators to monitor relatively minor changes in alertness.

The paradoxical combination of prolonged latency and increased amplitude, found in relation to evoked-response changes with age, is once again encountered in sleep data. Favale et al. (1964) obtained evidence that afferent transmission at the thalamic level is maximal during deep sleep in the cat. This facilitation of ascending impulses could explain increased amplitude of the early components of evoked responses during deep sleep. However, the prolonged latencies do not seem to be in accord with facilitation and suggest inhibitory mechanisms causing slowing of transmission. It thus appears that sleep either involves both facilitation and inhibition, or a combination of reduction and increase in inhibitory activity. The predominant effect could vary in different structures. Increased latency, representing slowed transmission of sensory information, seems more compatible with the functional changes accompanying sleep than increased amplitude. However, increased amplitude could also be an inhibitory effect, if it resulted from inhibition of cortical inhibitory activity.

DELIRIUM

Delirium generally refers to a disorder of the sensorium in which orientation is impaired, critical faculty is blunted or lost, and thought content is irrelevant and incoherent (Hinsie and Campbell, 1960). The acute and chronic brain syndromes may be regarded as deliria.

Delirium Tremens

Gross *et al.* (1964) studied four male, chronic-alcoholic patients admitted because of acute toxic complications. Evoked responses to clicks were recorded about 10 to 12 hr after the patients had received paraldehyde. Results suggested an association between the click-evoked response pattern and the presence of hallucinations. On days when subjects displayed hallucinations, an initial component, ordinarily peaking at 25 to 30 msec, appeared lower in amplitude, broader, and the latency of the peak was prolonged.

Experimental Delirium

Ditran® (Lakeside Laboratories) is an anticholinergic agent that regularly produces delirium in dosage of 10 mg intramuscularly (Meduna and Abood, 1959; Wilson and Shagass, 1964). The reaction is relatively stereotyped, reaching its peak about one hour after injection. It is characterized by: fluctuating unresponsiveness to questioning, with brief lucid intervals; rambling incoherent speech; severely impaired recent memory and attention; confabulation; disorientation; and abulia. Hallucinations, frequently of auditory type, are commonly reported. There is subsequent amnesia for most of the experience. The EEG changes with Ditran are like those found in other delirious states (Flugel and Itil, 1962), with increased slow-wave activity. The Ditran delirium is dramatically reversed by tetrahydroaminoacridine hydrochloride (THA) (Gershon and Olariu, 1960). With Ditran and THA, it is then possible to alter the sensorium and reverse the effects in a reliable manner, and we used these agents to study the accompanying changes in evoked potentials (Brown *et al.,* 1965).

Some indication of the disruption of cognitive functioning produced by Ditran may be obtained from the figure drawings in Figure 69. These were obtained in a comparative study of the behavioral effects of LSD and Ditran (Wilson and Shagass, 1964). The subject of Figure 69 was one of only five in a group of 11 who were able to draw at all under the influence of the drug. Somatosensory and visual evoked responses were recorded in 13 volunteer psychiatric patients before and after administration of Ditran in two divided doses of 5 mg each. The effects of the two doses were

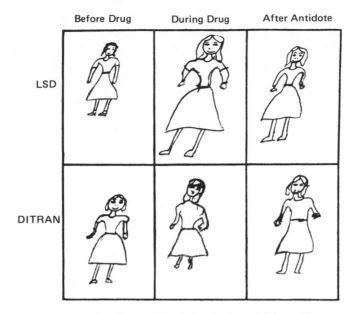

Fig. 69. Effects of LSD, 2.5 µg/kg iv, and Ditran, 10 mg, im, on figure drawing in the same subject. Drugs given one week apart. Subject was one of five in a total of eleven who could draw, at all, under influence of Ditran. (Reprinted with permission from Wilson and Shagass, 1964.)

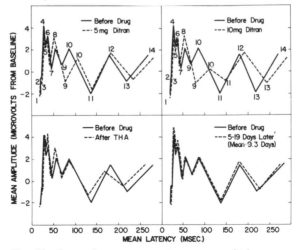

Fig. 70. Composite mean somatosensory evoked-response curves for 13 subjects, comparing predrug responses with those recorded under influence of Ditran, 5 and 10 mg, im, and after administration of Ditran antidote, tetrahydro-aminacridine (THA). Latencies of later waves were prolonged with Ditran and restored after THA. (Reprinted with permission from Brown *et al.,* 1965.)

additive. THA was then administered intravenously in doses of 30 to 60 mg.

Figure 70 shows mean measurements of somatosensory response amplitude and latency comparing the predrug records with those obtained after the two doses of Ditran, after THA was given, and with those obtained 5 to 19 days after the administration of Ditran. The main effects occurred in the later components numbered from 9 on. These were prolonged in latency. In contrast, there was a statistically significant decrease in latency of peak 4, which was accompanied by an amplitude decrease. Amplitude of peaks 8 and 9 increased. Changes in the later components of the somatosensory response resemble some of those observed in sleep (Figure 64). These changes in the later components were reversed by THA. The mean responses recorded several days after drug administration were almost identical with those obtained before the drug was given.

Figure 71 shows the mean measurements made for the first 300 msec of the evoked response to flash. Significant increases in the latency of components 7 and 8 and in the amount of negativity in components 3 and 5 were found on statistical analysis. Significant decreases occurred in the latencies of components 1 to 4 and in the amplitudes of peaks 4, 6, 9, and 11. The most striking effect on the visual response was the abolition of the after-rhythm. This occurred in all subjects and is illustrated in Figure 72. THA only partially reversed these effects on the visual response which were, however, no longer evident when the subjects were retested 5 to 19 days after drug administration.

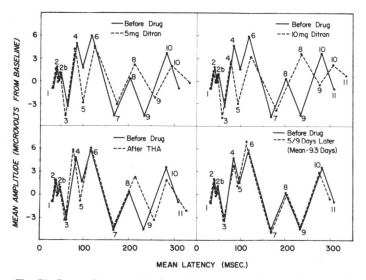

Fig. 71. Composite mean visual evoked-response curves in the same subjects as in Fig. 70 with Ditran and THA. Note latency prolongation of later waves with Ditran partially restored to predrug values after THA. (Reprinted with permission from Brown *et al.*, 1965.)

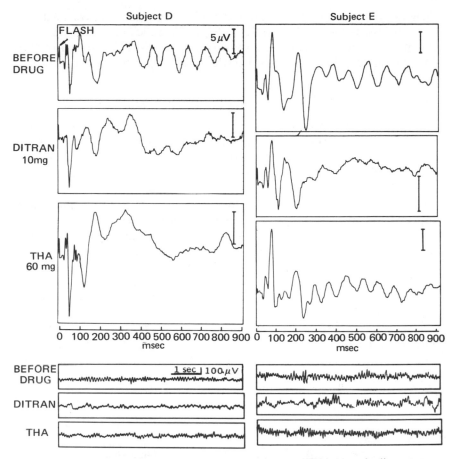

Fig. 72. Visual evoked responses of two subjects given Ditran and THA. Note the disappearance of after-rhythm in both subjects with Ditran. THA restored after-rhythm in subject *E*. EEG samples, at bottom, show increased slow activity and reduction of alpha activity with Ditran; also, restoration of normal pattern with THA. (Reprinted with permission from Brown *et al.*, 1965.)

The findings with Ditran suggested that the earlier portions of the responses were speeded by Ditran while the later portions were reduced or slowed. Since the pupils were not fixed, the shorter latencies of early components could have been due to Ditran-induced mydriasis. However, the lower amplitude of the flash-response early components and the speeding of the somatosensory early component seem to argue against such an effect. The main reversing effects of THA were on the later portions of the responses. Since THA quickly normalized the sensorium, this suggested that the later components were more specifically associated with processes related to sensorium than the early ones. The abolition of the after-rhythm in the visual response would be expected in association with the EEG changes, which consist of a combination of slowing and some fast activity. Nevertheless, as the EEG examples in Figure 72 indicate, ringing was often absent even when rhythmic activity was restored by THA and also when EEG rhythmic activity was recorded while under the influence of Ditran.

Chronic Delirium (Organic Brain Syndrome)

Cohn (1964) observed a simplified visual evoked response without after-rhythm in a patient with Jacob–Creutzfeldt syndrome. Straumanis *et al.* (1965) compared visual evoked responses in 20 patients with arteriosclerotic brain syndromes with those of 18 healthy control subjects. Mean ages of the patients and controls, respectively, were 74.4 and 71.8 years. The patients were characterized by disorientation in time and place, impaired judgment, and very poor performance on tests of memory and intelligence. The controls showed normal performance on intelligence and memory tests with no signs of brain syndrome. The clinical EEG was considered abnormal in 70 % of the patients and in 22 % of the controls.

Figure 73 shows the mean composite visual evoked responses of the patients and controls. Statistical analysis revealed that the amplitude of peak 2 was significantly greater in the patients and that latencies of peaks 6, 7, and 8 were significantly prolonged. A quantitative comparison showed much less after-rhythm in the patients than in the controls. The clinical EEG report classified the EEG as normal or abnormal and also classified alpha

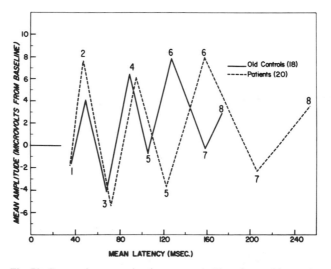

Fig. 73. Composite mean visual responses in 20 patients with arterio sclerotic brain syndrome and 18 healthy nonpatients matched for age and sex. Note prolonged latencies and somewhat greater amplitudes in patients. (Reprinted with permission from Straumanis *et al.*, 1965.)

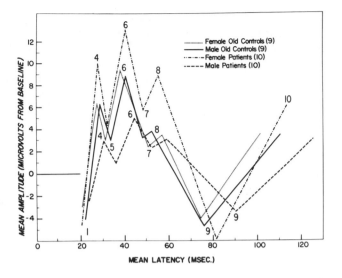

Fig. 74. Composite mean somatosensory responses in patients with arteriosclerotic brain syndrome and healthy nonpatients. Responses of male and female nonpatients do not differ but female patients have greater and male patients have lower amplitude responses than nonpatients.

activity as "good" or "poor." Although the amount of visual response afterrhythm was not related to the classification of the EEG along the normal–abnormal dimension, it was significantly related to the presence of "good" or "poor" alpha rhythm. As might be expected, after-rhythm was greater when alpha was "good."

Somatosensory evoked responses and their recovery functions were also studied in the same group of brain syndrome patients. The composite somatosensory responses, shown in Figure 74, summarize the results. The main finding was a large difference between the sexes in the patient group, whereas no such difference was evident in the healthy controls. Statistical analysis showed that the amplitudes of peaks 1, 4, and 6 were greater in female than in male patients. The shorter latencies of the responses in females were as ordinarily found. The amplitude differences by sex in patients were further supported by analysis of variance which revealed significant interactions between sex and diagnosis for the peak 4 measurement and the peak-to-peak measurement from 1 to 4.

Recovery function results showed only that the initial negative peak of the somatosensory response recovered more in the brain syndrome group than in the controls. There were, however, significant interactions between sex and diagnosis for the adjusted R_2 amplitudes of peaks 4, 7, 9, and the adjusted R_2 latency of peak 8. In addition, variations in the shape of the recovery curve were revealed by significant triple interactions with sex and diagnosis for amplitudes 6, 8, and 9 and latency 8. These findings suggest that the somatosensory response and its recovery functions are differentially affected in males and females in the presence of arteriosclerotic brain syndrome. The only finding not complicated by the sex factor was the greater recovery of peak 1 in the brain syndrome patients. Although this would suggest "hyperexcitability" in the patients, the finding should be viewed with caution since it could have occurred by chance as one of a large number of statistical comparisons. Also, the recovery of peak 4 was actually less in the patients, although not significantly so, whereas the same diagnostic difference might have been expected for the positive as for the negative phase of the initial component.

In a recent study, Malerstein and Callaway (1969) applied the two-tone averaged evoked response procedure to patients with Korsakoff's syndrome. The two-tone evoked response measure will be considered further later in relation to studies of schizophrenia. Its essence is that it measures the similarity between responses to two tones of different frequency, which are represented in a pseudorandom series. The degree of similarity is estimated by computing the correlation between the values summed in successive memory locations of the computer. Malerstein and Callaway studied 23 patients with the diagnosis of chronic brain syndrome, alcoholism. The similarity scores of these Korsakoff patients deviated from normal and were more like those found by Callaway et al. (1965) in schizophrenic patients. Furthermore, there was a tendency for the scores to be correlated with clinical ratings of memory impairment. Greater evoked-response similarity, i.e., resemblance to normal, was associated with less impairment.

Malerstein and Callaway also studied eight patients with senile psychosis. The results in the seniles were essentially similar to those of the Korsakoff patients. However, in the senile patients, there was no correlation between degree of memory defect and the two-tone score, as had been found in the Korsakoff group. Malerstein and Callaway interpret this lack of correlation to suggest that memory defect in itself does not produce the correlation found between clinical impairment and the evoked-response score in Korsakoff patients.

COMA

Bergamasco et al. (1966) made serial records of visual responses in patients with post-traumatic coma unaccompanied by clinical signs of localized lesions. The pupils were fixed in mydriasis by application of atropine. The visual response was found to be markedly altered in coma when the EEG was predominantly occupied by 2 to 3 c/sec activity. Response latency was increased to about 100 msec and the response was composed of only two or three phases. The after-rhythm was not present. Recordings made between frontal and vertex electrodes showed no activity; a response was recorded only from the occipital-vertex derivation. When the depth of coma was progressively decreasing and EEG activity was faster, the response from the occiput was more organized and contained waves of shorter latency and duration. Also, at this stage, some evoked activity was recorded in the frontal areas.

Figure 75 illustrates the changes observed in patients who gradually returned to consciousness; the evoked responses showed shortening of latency and increasing complexity of wave form. However, even when consciousness was regained and alpha activity was prevalent in the EEG, Bergamasco *et al.* did not observe the after-rhythm. In a patient who went on to death, the evoked response shortly before death was further reduced in amplitude.

Cohn (1964) also observed absence of after-rhythm in patients in semi-coma resulting from a subdural hematoma. This author shows records

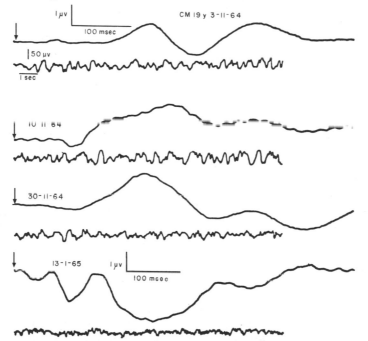

Fig. 75. Visually evoked potentials (occipital, 100 responses, 1 per second) in a patient from coma to arousal. Representative EEG samples underneath each average response. Top line, coma: the evoked potential is composed of three phases beginning about 150 msec after the stimulus; continuous delta in EEG. With lightening of coma, there is a slight increase in amplitude of the waves and shortening of the latencies in the two central recordings, and EEG activity is faster. Bottom tracing: the patient was conscious and the evoked potential was almost normal, while the EEG showed a discontinuous alpha activity. After-rhythm was not seen. (Reprinted with permission from Bergamasco *et al.,* 1966.)

of a case of coma resulting from cerebral anoxia following cardiac arrest, which resemble the records of Bergamasco *et al.* Lille *et al.* (1968) studied evoked responses in comatose children who were mainly between 1 and 6 years of age. Responses to flash, click, and median shock all showed the same general characteristics in relation to depth of coma. The wave forms were less complex during coma, containing fewer components. They were also of lower amplitude and longer latency. Lille *et al.* considered the characteristic most related to depth of coma to be the complexity of the wave form.

Lille *et al.* related the characteristics of evoked responses to those of the EEG. The EEG's were classified as: normal; slow-wave; low-voltage; paroxysmal. Evoked responses associated with low-voltage EEG's had the lowest amplitude, least complexity, and longest latency. Amplitude was also diminished in responses obtained in association with paroxysmal EEG's. The responses with slow-wave EEG's were intermediate between those obtained with normal and low-voltage EEG's. Responses recorded when the children were having convulsions were of reduced amplitude during the attack and increased in amplitude after it terminated. Although Lille *et al.* found some relation between evoked-response characteristics and the clinical evolution of the case, they were unable to establish any correlations with the etiology of the coma.

The observations of Lille *et al.*, during seizures, appear to agree with those obtained by Corletto *et al.* (1966), who found that the evoked response became flattened during convulsions and that a postictal augmentation of amplitude took place. However, the Lille results in childhood coma seem to be somewhat discordant with the emphasis placed by Bergamasco *et al.* (1966) on the correlation in coma between evoked-response modification and the presence of delta activity in the EEG. Lille *et al.* found that the evoked-response changes, in the presence of a flat EEG, were more marked than those obtained when the EEG was characterized by slow-wave activity.

COMMENT

The evoked-response changes in sleep, delirium, and coma appear to be similar in several respects. All of these conditions are associated with prolonged latency of later components, abolition of after-rhythms in the visual response, and simplification of wave forms. However, there are also notable differences. In the acute delirium induced by Ditran, but not in sleep, chronic delirium, or coma, the initial components of the responses were either unchanged in latency or actually speeded. Tendencies toward augmentation in amplitude of early components were observed during sleep and in chronic brain syndromes. In contrast, the early components were either obliterated or greatly slowed and reduced in amplitude in comatose states. It appears, therefore, that the later events in evoked responses undergo alterations of a similar nature in all states involving major impairment of consciousness, whereas the changes in the earliest components may differ quite markedly from one state to another.

One speculative interpretation of the changes in the early components is that they are related to the potential for being roused. Persons in a deep sleep or in a delirium are responsive to external stimuli, whereas a patient in coma is not. This suggests that the presence of well-defined early evoked-response components may be associated with the capacity for responding with some type of coordinated action to external stimulation, and that their absence may signify the loss of such a capacity. Although attractive, this interpretation seems discordant with the persistence of early components in surgical anesthesia (Domino and Corssen, 1964). However, anesthesia may involve special features, since it is quickly induced by external agents and quite rapidly reversible.

It seems reasonable to suppose that the earliest evoked-response events are related to transmission of information in the CNS and that the later ones may reflect activities concerned with information processing. The fact that the later events are altered in all states of impaired consciousness seems to agree with this supposition. Furthermore, the nature of the evoked-response changes, particularly the marked slowing, seems concordant with the manifest reduction of information-processing activities in these states.

Chapter 6

Attention and Related Phenomena

Attention implies the direction of the mind to an object, concentration on particular stimuli in a complex, or the state of consciousness characterized by such concentration. Evoked-potential correlates of attention and related phenomena such as vigilance, distraction, and habituation have been explored in numerous studies. The recording of some kinds of event-related potentials, such as the CNV, is contingent upon employing experimental paradigms in which aspects of attention are manipulated (Chapter 3). Indeed, few recording situations can be considered to be free of the influence of variables related to attention. This chapter presents material selected to illustrate some main findings concerning evoked potential correlates of attention. The reader is referred to reviews by Callaway (1966), Näätänen (1967), Tecce (1970), and to the symposium reports edited by Dargent and Dongier (1969) and Donchin and Lindsley (1969) for additional material.

HABITUATION

Habituation is said to occur when response size diminishes with repetitive stimulation. Perry and Childers (1969) suggest that habituation may be distinguished from fatigue by its more transitory nature since reversal of habituation will take place if novelty is introduced or if the stimulation sequence is briefly interrupted; such reversal does not characterize fatigue. Walter (1964a) has put forward the idea that habituation may occur when a "threshold of triviality" is attained; according to this, habituation may represent the opposite of attention.

Two kinds of habituation may occur during evoked-response recording. There may be habituation *within* a single averaging sequence, or response decrement may occur *between* sequences. Special analytic procedures

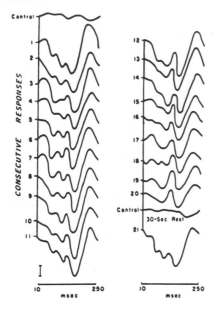

Fig. 76. Consecutively recorded visual evoked responses displaying habituation. Pupil was dilated. Electrodes on inion and 5 cm anterior in midline. The light was flashed but occluded for control trials. Note progressive reduction of amplitude in consecutive averages with some variability and increase of amplitude after 30 sec rest. (Reprinted with permission from Perry and Copenhaver, 1965.)

are required to demonstrate within-sequence habituation (Ritter *et al.*, 1968). Between-sequence habituation is demonstrable by displaying successive averages as in Figure 76; this figure also illustrates the waxing and waning of amplitude during a long sequence of recording, even though the general trend is toward decrease.

The relative role of central and peripheral factors in evoked-response habituation has become a subject of controversy. Bergamini *et al.* (1965) have shown that habituation of the visual response occurs in 10 min when the pupils are mobile, but that there is no evidence of habituation in 30 min when the pupils are fixed. They also report absence of habituation in subjects with congenital absence of the iris (Figure 17), and delayed habituation in patients with Adie's syndrome, in which a state of pupillary tonicity permits a pupil reaction to light only after a lengthy period of illumination (Bergamini and Bergamasco, 1967). These observations suggest that progressive miosis is responsible for habituation of the visual response and that habituation does not occur when the amount of light energy on the retina is kept constant. On the other hand, habituation of the visual response has been demonstrated by other workers who have kept pupil size constant either by pharmacologic fixation, use of an artificial pupil, or by employing subjects with congenital aniridia and complete ocular paralysis (Garcia-Austt *et al.*, 1963; Perry and Copenhaver, 1965). Tecce (1970) points out that the conflicting results were obtained in experiments which differed in a number of procedural factors such as stimulus rate, distance from the eye, monocular *vs.* binocular stimulation and electrode placement. Another factor which may be of considerable importance is the previous experience of the subject. Perry and Childers (1969) are of the opinion that habituation rarely occurs when the subject is being used for the first time, whereas it is common in trained subjects.

Habituation has been reported for auditory and somatosensory responses as well as visual (Allison, 1962; Bogacz et al., 1962; Cigánek, 1965; Uttal, 1965). Assuming that the stimulating arrangements remain stable, it is difficult to attribute evidence of habituation in the somatosensory response to peripheral factors.

In our own studies of visual and somatosensory responses, we obtained some evidence of response decrement and latency increase that could be interpreted as habituation. During recovery-function determinations, the responses to unpaired flashes decreased significantly in amplitude (Shagass and Schwartz, 1965a). However, paradoxically, latency of the initial negative deflection also decreased; this is not consonant with habituation. While measuring somatosensory recovery functions, the amplitude of several R_1 peaks decreased and the latencies increased significantly over the course of the test session (Shagass, 1968a). Although statistically significant, the magnitude of the habituation changes were relatively slight. The fact that these observations were made during recovery-function measurement also casts some doubt upon the interpretation of the results as habituation. Since the interval between paired stimuli was systematically lengthened as the recording session progressed, it is possible that the greater proximity of the second stimulus to the succeeding unpaired stimulus was responsible for the observed response decrement.

Perry and Childers (1969) outline a number of procedures to minimize habituation. They include: frequent rest periods, counterbalancing of stimuli, aperiodic rates of stimulation, and exclusion of the first few individual responses from analysis. On the other hand, if one wishes to study habituation as a phenomenon in itself, the opposite recommendations can be made. Under the usual recording conditions, there is probably much variability in the degree to which habituation is favored. Rest periods may be longer, interruptions more frequent, and conversation between runs more often undertaken with some subjects than with others. Conclusions about differential susceptibility to habituation in patient and control populations are difficult to reach under these circumstances. In one of our studies (Shagass and Schwartz, 1965a) we obtained results suggesting more habituation in patients than in nonpatients, but these findings are suspect because no attempt had been made to control the factors influencing habituation between recording sequences.

DIRECTED ATTENTION

Experimenters have attempted to direct attention by verbal instructions, such as telling the subject to count stimuli, to concentrate on stimuli with certain qualities, and to disregard others, to respond to some stimuli by pressing a key, to make discriminations between weaker and stronger stimuli, etc.

Counting Stimuli

Garcia-Austt and his coworkers found that counting repetitive light flashes augmented the visual response (Garcia-Austt, 1963; Garcia-Austt et al., (1964). Williams et al. (1964) and Gross et al. (1966) observed auditory responses to be increased by counting. However, evoked-response amplitude does not always increase with counting. Spong et al. (1965) reported that only 6 of 13 subjects showed the phenomenon. Unpublished data obtained in our own laboratory showed both increases and decreases in visual-response amplitude when subjects counted flashes, compared to when they did not.

Tecce (1970) suggests that some of the variability in counting results may be due to variations in fixation of the eyes, since Gastaut and Regis (1965) observed enhancement by merely asking the subject to look attentively at the flashing stroboscope. Spong et al. (1965) have suggested that counting may involve distraction from the stimulus while keeping track of the count. Subjective reports of subjects participating in counting experiments suggest that the relative ease and automatic nature of the counting performance may be important; many subjects state that counting stimuli involves no particular effort or need for special direction of attention.

Differential Attention

Spong et al. (1965), in addition to requiring their subjects to count stimuli in a particular sensory modality, also asked them to perform a vigilance task. The subject was instructed to press a key after detecting an occasional weaker or dimmer stimulus that occurred among the more numerous louder or brighter auditory and visual stimuli. With visual vigilance, the subject attended to flashes, ignored clicks and was required to press a key when occasional dim flashes occurred. When engaged in the auditory vigilance task, he was required to press the key in response to occasional weak clicks interspersed among the more numerous louder clicks. Stimuli in the modality other than that for which the instructions demanded attention were ignored. Figure 77 illustrates the results obtained by Spong et al. When the subject was attending to flashes in the vigilance task, these elicited larger responses than the same flashes when he was attending to clicks. Similarly, the click responses were of greater amplitude when attention was directed toward them. The data were similar during key pressing.

Satterfield and Cheatum (1964) administered shock stimuli to the

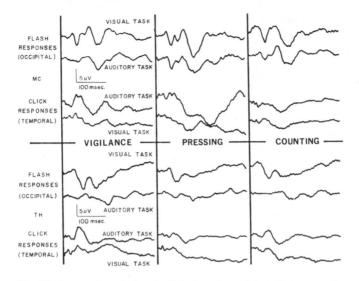

Fig. 77. Averaged evoked potentials from two subjects in response to flashes and clicks. Subjects performed visual and auditory tasks under three experimental conditions: vigilance, key pressing, and counting. Note increased amplitude of responses in sensory modality involved in the task. (Reprinted with permission from Haider, 1967.)

right and left wrists and instructed the subject to ignore the stimuli delivered to one wrist and attend to those delivered to the other. The responses to the nonattended shocks were unchanged while those elicited by the attended shocks were generally increased, although they were reduced in 5 of 25 subjects. Satterfield (1965), in another experiment, presented clicks and shocks alternately and asked the subjects to attend to one or other stimulus. Figure 78 illustrates the large differences obtained in several subjects as a function of the direction of attention. A late response component, peaking at about 150 msec in vertex electrodes, was markedly augmented when the stimulus was the focus of attention. It is noteworthy that the initial components of the response to shock were not affected by the instructions concerning attention. Satterfield also recorded the peripheral nerve response in a few subjects and found no changes in relation to attention. There were a few subjects (seven of 47) who did not follow the general trend of increased amplitude of the late components with attention.

Chapman and Bragdon (1964) found that the visual response to number stimuli was larger than to nonnumber stimuli, to which attention was not required, when the subjects were asked to make a judgment about the relative magnitude of numbers. Chapman (1965), in another experiment, showed that the finding was independent of the light energy content of the stimuli. Shevrin and Rennick (1967) showed that attention directed to the stimulus augmented the late components of responses to tactile

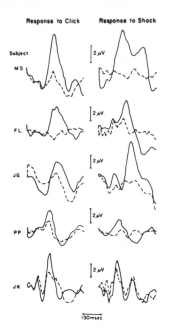

Response to Click   Response to Shock

Subject
M S

F L

J G

P P

J K

130msec

Fig. 78. Average responses to click and to shock for one trial in each of five subjects. Solid lines indicate subject attending to stimulus; dotted lines indicate subject not attending to stimulus. (Reprinted with permission from Satterfield, 1965.)

stimuli when compared with those obtained when the subject was free associating or doing mental arithmetic.

"Unconscious" Attention

Shevrin and Fritzler (1968) have obtained evidence that visual stimuli that are not consciously perceived may produce differential evoked-response effects. They employed two stimuli matched for size, configuration, and color but differing in meaning. The meaningful stimulus was a picture of a pen and a knee. The abstract one was composed of figures that approximated pen and knee in size, shape, and color but lacked the distinctive contour features of the real objects. When these pictures were flashed for 1 msec the subjects were unable to describe their content. However, evoked responses recorded between occiput and vertex differed quantitatively. The amplitude of an occiput-positive component, peaking at approximately 160 msec, discriminated between the meaningful and abstract stimulus. Free associations collected during the course of the experiment gave evidence that the meaningful stimulus had influenced mental activity. This suggests that the evoked response can discriminate between two "subliminal" stimuli with differential effects on subsequent thinking.

Reaction Time

Fast reaction time may be interpreted as evidence of a high level of attentiveness. Several investigators have found that visual response amplitude measured during a reaction-time experiment was greater in association with shorter reaction times to flash (Dustman and Beck, 1965b; Donchin and Lindsley, 1966; Morrell and Morrell, 1966). Figure 79 illustrates the variations in flash-evoked responses associated with different reaction times. Eason *et al* (1967a) found that longer reaction time was related to decreased visual response amplitude if the latter was associated with decreased light intensity. Eason and White (1967) found no relationship between visual response and reaction time in one subject. Wilkinson and Morlock (1967) obtained equivocal results for the relationship between auditory-response amplitude and reaction time. A recent experiment in our laboratory (unpublished observations) in which reaction time to flash was measured, also yielded evidence supporting increased amplitude in association with faster reaction times. However, several subjects showed an opposite trend.

On the whole, then, the evidence suggests that the direction of attention to a stimulus tends to result in augmentation of evoked-response amplitude. The augmentation occurs in response components that peak at about 100 msec or later and does not seem to be evident in the earliest components. It is noteworthy that the augmentation of amplitude of late components with directed attention is not universal and that some subjects may show decreases. However, many of the exceptions may be explained in terms of the failure of verbal instructions to produce the desired effects on attentive behavior.

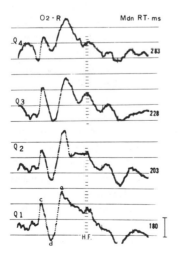

Fig. 79. Differences in averaged photic evoked responses (O_2 to linked ears) arranged according to speed of reaction time. Each tracing represents average of 62 trials. Sweep duration 512 msec, calibration 10 μv. (Reprinted with permission from Morrell and Morrell, 1966.)

Do Evoked Responses Reflect Selective Attention?

Näätänen (1967) has criticized most of the experiments cited above from the viewpoint that their results do not provide conclusive evidence for *selective* attention. In his view, evoked-response evidence of selective attention should demonstrate that the stimulus to which attention is directed gives an enhanced response while the responses to stimuli not attended to are diminished. Näätänen emphasized the procedural point that previous experiments in selective attention had involved alternating stimuli so that subjects had foreknowledge of the stimulus to come. He conducted three experiments which attempted to overcome the deficiencies he had perceived in other work.

In one experiment, Näätänen measured responses evoked by clicks during visual attention and during relaxation. He found that click-evoked responses were augmented during visual attention and took this to indicate that changes in brain reactivity associated with attention influence responses in all modalities in the same direction, either enhancing or depressing their amplitudes. This would mean absence of selectivity. In a second experiment, he required subjects to discriminate between relevant and irrelevant stimuli which were presented in a randomized order and at varied intervals. He found no statistically significant differences between the responses to relevant and irrelevant clicks, even though the subjects discriminated well. In the third experiment, Näätänen demonstrated increased amplitude with relevant stimuli when they were presented *regularly* in alternation with irrelevant stimuli. However, in the same subjects a randomized order failed to yield differences in the responses. Näätänen attributed the enhancement with regularly presented stimuli to anticipatory and preparatory arousal and activation preceding relevant stimuli.

The recent findings of Eason *et al.* (1969) do not, however, lead to conclusions identical with those of Näätänen. Eason *et al.* had subjects attend to sharply focused, low-intensity visual stimuli in one visual field by requiring a reaction-time response. Stimuli to the other visual field were "irrelevant." The timing of stimulus presentation was such that subjects could not predict the forthcoming stimulus. Arousal was manipulated by introducing a condition of threatened painful shock to the foot in some trials. The results showed much larger responses to the relevant stimuli. There were often no discernible responses to the irrelevant ones. A control run without manual responses indicated that the results were not dependent on the reaction-time feature of the experiment. Responses were also greater, and reaction times were shorter, under threat of shock than under no threat; with shock threat, the differences in response to relevant and irrelevant stimuli were magnified. The data of Eason *et al.* appear to provide clear evidence that selective attention is reflected in evoked responses. Their results also indicate that arousal increases response amplitude; this is in agreement with Näätänen's observations. The use of low-intensity stimuli in the same sensory modality by Eason *et al.,* which contrasts with the cross-modality comparisons, usually employing stronger stimuli, of other investigators, may have contributed significantly to the experimental outcome.

HYPNOSIS AND SUGGESTION

Hypnosis may be regarded as a special attentive state. Attention is focused on the hypnotist's voice and can be especially well controlled by the hypnotist's instructions. Hernandez-Péon and Donoso (1959) reported that visually evoked responses recorded from leads implanted in the white matter of the occipital cortex of two patients were increased or decreased depending upon the suggestion of higher or lower light intensity. Guerrero-Figueroa and Heath (1964) also observed variations of the visual response with nonhypnotic suggestion in an implanted patient. Clynes *et al.* (1964) found that the hypnotic suggestion of blindness reduced the visual response in one of two subjects.

In our laboratory, we attempted to determine whether the somatosensory response was reduced by hypnotically induced anesthesia of the stimulated arm in one subject (Shagass and Schwartz, 1964*a*). Even though, under hypnosis, the subject disclaimed perception of pain when his skin was clamped with a hemostat, no change in the evoked response was observed. However, later questioning suggested that he may not have been anesthetic to the shock stimulus used to produce the response. We also attempted to determine whether the hypnotic trance, *per se,* alters the early somatosensory recovery function. Recovery was measured over several interstimulus intervals to 50 msec before and after induction of hypnosis in 12 experienced nonpatient hypnotic subjects. With the first several subjects, it seemed that the initial recovery phase was being augmented by hypnosis. However, it turned out that, although still responsive to the hypnotist's command, the subjects were actually dozing. When drowsiness was prevented, the changes in the recovery function measurements were no longer observed and there was no significant change in amplitude of response to the first stimulus.

Halliday and Mason (1964) found no reduction in the somatosensory response in a group of subjects who reported little or no perception of the stimulus during hypnotic anesthesia. Figure 16 illustrates some of their results. These authors also found that the auditory response to click did not change during attempts to induce hypnotic deafness; however, they did not obtain complete deafness to the clicks. Beck and Barolin (1965) and Beck *et al.* (1966) were unable to find changes in the visual response to hypnotic suggestions of stimulus brightness and dimness in a group of subjects.

As Tecce (1970) points out, the evidence for evoked-response changes with hypnosis has been negative whenever more than two individuals have been studied. However, the possible influence of hypnosis on evoked-response phenomena cannot be regarded as a closed issue. Clinicians who practice hypnosis draw attention to wide individual variations in responses to hypnotic suggestion and to the existence of extreme responders who are capable of undergoing major surgical operations without anesthesia. Evoked-response investigation of this special group would seem worthwhile.

DISTRACTION

Various techniques have been employed to direct the subject's attention away from the stimulus. The common one is to require mental activity during stimulation. Cigánek (1964) found that the visual response was smaller during mental arithmetic than during counting of stimulus flashes or during a discrimination task (Cigánek, 1967). He believed that the effects of attention were more prominent in responses to paired than to single flashes. Garcia-Austt et al. (1964) noted that the visual response was reduced when distracting tones or clicks were introduced, as well as by mental arithmetic. Shevrin and Rennick (1967) found reduction of the response to touch during mental arithmetic and free associations. Guerrero-Figueroa and Heath (1964) reported that hippocampal responses evoked by central stimulation were reduced by mental calculation, and Ervin and Mark (1964) found a similar reduction for a tactile response in an implanted patient.

The foregoing results would suggest that distraction regularly reduces the evoked response, but there are contradictory findings. Eason et al. (1964) found augmented visual responses with mental arithmetic and greater increases for tasks requiring more mental effort. They also found increased responses when the subjects thought about visual, auditory, or kinesthetic stimuli, memorized digits, or recited the alphabet. Eason et al. consider their data to support the idea that increased arousal enhances the visual response, but their evidence also supports the conclusion that response amplitude is increased by distraction. Tecce (1970) points out that the experimenters served as their own subjects and the discrepancy between their results and those of others may result from this factor. Tecce also described some of his own results which showed increased amplitude of visual responses during distraction. On the other hand, random shock presented during photic stimulation reduced the visual response.

In a recently completed study in our laboratory (Shagass et al., 1970) somatosensory recovery functions were recorded during conditions of "rest" with eyes shut and while watching television (TV). The main purpose of the study was to determine whether differences between psychiatric patients and nonpatients in somatosensory recovery functions can be attributed to variations in state of attentiveness or distraction. Figure 80 shows the mean amplitude and latency measurements to unpaired shock

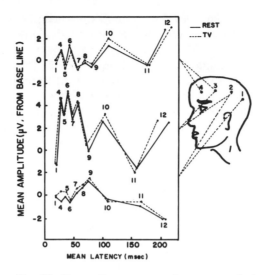

Fig. 80. Composite mean somatosensory evoked responses from ten subjects recorded during conditions of rest with eyes shut and while watching television. Differences between two conditions minimal.

stimuli for ten subjects. No difference between "rest" and TV conditions is apparent. Neither were significant recovery-function differences found using R_2 values adjusted for their covariation with R_1. Concurrent EEG measurements did reveal marked amplitude reductions during TV viewing, but these changes in EEG amplitude were not significantly correlated with any evoked-response changes.

It appears that distraction has variable effects upon evoked-response characteristics, sometimes increasing and sometimes decreasing amplitude. Explanation of the variable results will require better specification of the experimental conditions and the actual psychological states engendered by them.

LONG-LATENCY POTENTIALS

Positive response components with latencies of 300 to 500 msec, which appear to be related to complex psychological attributes of the stimulus, have been described in Chapter 3. Some of the reports dealing with evoked-response correlates of attention may have involved measurements of these components without specifically separating them from earlier ones. Sutton (1969) has provided a comprehensive and thoughtful discussion of the conceptual and methodological problems associated with investigation of the correlates of the P_3 phenomenon. Incidentally, he drew attention to the fact that Näätänen (1967) had actually obtained positive data showing that task-relevant stimuli gave larger P_3 amplitudes, but that he had discounted these results.

Sutton's basic finding was that the amplitude of the P_3 wave was inversely related to the probability of occurrence of the stimulus. He, therefore, related the electrical event to the resolution of stimulus uncertainty. However, Teuting (1968) showed that, although P_3 was markedly attenuated when the subject knew in advance what the stimulus would be, its amplitude still varied as a function of the probability that a click of given intensity would occur. These results suggest that the notion of stimulus uncertainty does not adequately encompass the phenomenon. Sutton also gives reasons not to accept the term "significance" to designate the correlated psychological phenomenon, and he discounts the applicability of the concept of generalized arousal. Another concept that has been proposed is "task-relevance." However, this does not accord with the data of Ritter et al. (1968). These workers found large shifts in P_3 when tones were changed to a different frequency while subjects were reading a book and presumably paying no attention to the tones. Sutton expressed the opinion that understanding of the P_3 component in terms of orienting or dishabituation may offer the most fruitful possibilities. This opinion is in accord with the view of Ritter et al. (1968).

Ritter and Vaughan (1969) recently attempted an assessment of the psychological significance of the late positive component. Figure 81 reproduces some of their data. Subjects were required to make auditory and visual discriminations. Vertex and occipital leads referenced to the chin revealed late-positive components for detected signals. These waves were absent for undetected signals or for nonsignals. Ritter and Vaughan also recorded with bipolar leads from vertex to occiput in order to reproduce the recording conditions of Haider et al. (1964). The late positive component did not occur in the bipolar derivation, since it was apparently common to both leads and was, therefore, cancelled out.

Ritter and Vaughan report several experiments conducted to explore various explanations for the results. Their evidence indicated that P_3 is not a motor potential. They also found that, when the discrimination between signal and nonsignal was made very difficult, P_3 was found for all stimuli whether these were detected or not. This finding indicated that the wave could not be a correlate of discriminating a change in stimulation. Another finding was that the P_3 wave tended to disappear in a reaction-time experiment which involved an easy discrimination.

Ritter and Vaughan discussed the possibility that the P_3 wave represents the comparator mechanism, presumably involved in the orienting situation, which must compare incoming stimuli with previously stored information. However, since P_3 was absent for nonsignals they concluded that it cannot represent activity of the comparator mechanism itself. Furthermore, since it was present for all stimuli in a difficult discrimination, it could not be

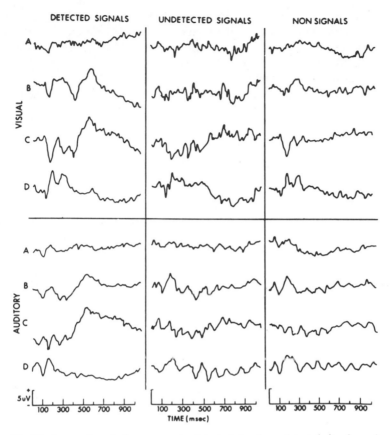

Fig. 81. Averaged evoked responses to detected signals, undetected signals, and nonsignals for one subject. Electrode placements: *A,* left supraorbital to chin; *B,* vertex to chin; *C,* right occipital to chin; *D,* vertex to right occipital. Trace begins with stimulus onset. In the visual condition, the averages are based on 33 detected signals, 11 undetected signals, and 33 nonsignals. In the auditory condition, the averages are based on 62 detected signals, 22 undetected signals, and 62 nonsignals. Averages for nonsignals were obtained by randomly sampling from each condition. Note large late waves in *B* and *C* with detected signals. (Reprinted with permission from Ritter and Vaughan, 1969.)

a correlate of conscious detection of a mismatch. Their final conclusion was that P_3 appears to be a correlate of central processes for cognitive evaluation of stimulus significance.

CONTINGENT NEGATIVE VARIATION (CNV)

The CNV phenomenon has been related to psychological events identified as expectancy, decision (Walter, 1964*a*), motivation (Irwin *et al.,*

1966; Cant and Bickford, 1967), volition (McAdam *et al.*, 1966), preparatory set or conation (Low *et al.*, 1966), and arousal (McAdam, 1969). Tecce (1970) points out that one state accompanying all of these is heightened attentiveness, usually to the imperative stimulus. Although the exact psychological correlate will depend upon the nature of the salient variables in any particular experimental situation, the important point in the present context is that CNV is demonstrated under conditions that can be construed as modifying the attentive state of the subject.

Variations with Concentration and Distraction

McCallum and Walter (1968) reported that the amplitude of CNV was greater during trials in which the subject reported himself to be concentrating than when he was not concentrating (Figure 82). Conversely, these authors and Tecce and Scheff (1969) have shown that the introduction of distracting stimuli between the alerting and imperative stimuli tends to reduce the amplitude of CNV (Figure 83). Occasional failure to reduce the CNV with distracting stimuli may be explained by the inadequacy of the distraction. McCallum and Walter (1968) reported results from a subject whose CNV diminished relatively little when spoken to, but who informed the experimenter that classical music would really attract his attention. In fact, when music was played, the magnitude of the CNV was markedly reduced.

Another line of evidence that CNV magnitude is increased by concentration comes from its variation with anticipated energy output. Low and McSherry (1968) required subjects to press a lever to three different pressures ranging from 2.5 to 10.5 kg and found that CNV was greater when the subject anticipated stronger pressure. On the other hand, CNV appears to diminish in association with fatigue and, presumably, the accompanying impairment of attention (Tecce, 1970; Dongier, 1969).

Sensory Discrimination and Multiple Anticipation

Low *et al.* (1967) conducted experiments in which the imperative stimulus was a click whose intensity was varied around sensory threshold. The magnitude of the CNV was determined by means of an area measurement and the variance of this measurement was also measured. The results clearly showed that, as the intensity of the imperative stimulus approached threshold, CNV magnitude increased and its variance diminished. These findings suggest that a difficult discrimination leads to a larger and more stable CNV. A troublesome aspect of the results lies in their apparent contradiction of the nonspecific nature of CNV, which has been supported by numerous experiments in which no particular variations were seen with either stimulus modality or intensity. However, Low's findings do seem consistent with the idea that CNV magnitude reflects the degree of "engage-

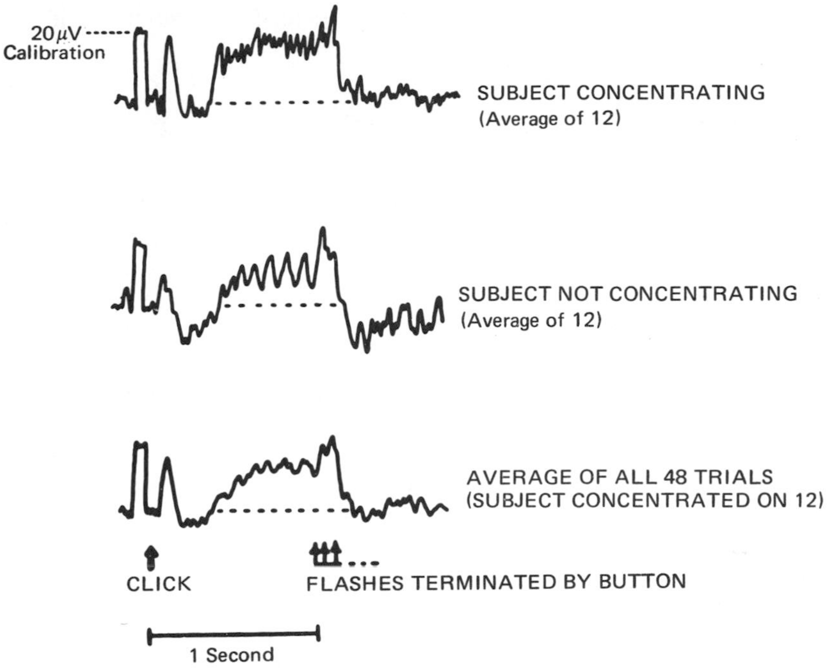

Fig. 82. Averaged CNV of a nonpatient subject in a series of 48 trials during which he voluntarily concentrated on some trials and not on others. Top trace is the average of 12 trials on which he concentrated. The second trace is the average of 12 trials taken at random on which there was no special concentration. Bottom trace shows the average of all 48 trials. Note the more rapid rise at onset and greater amplitude of CNV in the concentration trials. (Reprinted with permission from McCallum and Walter, 1968.)

ment" of the subject in activity related to the stimulus, since the task of discriminating at sensory threshold would involve greater difficulty.

Another experiment reported by Low and McSherry (1968) involved complicating the task for the subject by demanding simultaneous responses to two different reaction-time situations by pressing buttons with both hands. The CNV in the complex situation was greater than in either of the constituent situations alone. Low's data indicate that CNV can be markedly augmented in a situation of multiple anticipation so that the apparent maximal response level in the usual recording situation (Knott and Irwin, 1968) is mainly a function of task requirements. Evidence that only a small portion of frontal cortex, perhaps 2 % of available responding tissue, is ordinarily involved in CNV, has been obtained by Walter (1969) in his observations of patients with implanted electrodes. This would suggest that much more neural involvement can be obtained under the appropriate conditions of complexity and stress and this would result in larger CNV's.

Reaction Time

Walter (1969) noted that, during CNV studies, reaction time to the imperative stimulus may become extremely short. For example, times of 80 to 100 msec to a flash stimulus are not unusual, even though these values are well below the ordinary reaction time expected with visual stimuli. Hillyard (1969b) has observed intervals as low as 3 msec between imperative stimulus and button press. Clearly, there may be a large element of anticipatory reaction in the usual CNV situation involving precise timing. Such anticipatory reactions differ from the usual reaction time measurements, because they involve learning and depend upon the cue given by the alerting stimulus. Nevertheless, they may reasonably be taken to reflect a state of heightened attentiveness. Consequently, evidence that fast reaction times are associated with greater amplitude of CNV (Hillyard, 1969c) supports the relation between CNV magnitude and attentiveness. Figure 84 shows the CNV's associated with different reaction times. It should be noted, however, that systematic intraindividual relationships between CNV amplitude and reaction time, as in Figure 84, were observed in only half the subjects.

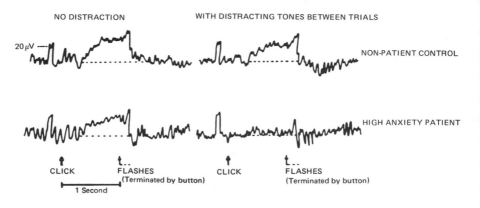

Fig. 83. Typical records of nonpatient control and high-anxiety patient in the acquisition and distraction conditions during CNV recording. The nonpatient retains his CNV quite well during distraction, whereas it is completely abolished in the high-anxiety patient. The click-evoked response is markedly attenuated in both subjects, a feature dependent upon the distraction (tone) being in the same sensory modality as the alerting stimulus (click). (Reprinted with permission from McCallum and Walter, 1968.)

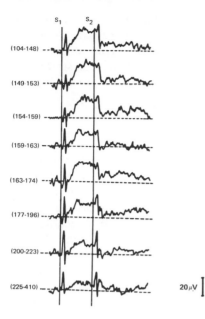

Fig. 84. CNV tracings, each the average from 12 trials, having the range of reaction times shown at left in msec. Negative up. Note that CNV amplitude preceding the motor response decreases as reaction time lengthens. (Reprinted with permission from Hillyard, 1969c.)

Time Estimation

Experiments have been conducted in which the interval between the alerting and imperative stimuli was systematically varied. When the interval was made quite long, the occurrence of the CNV was delayed; it generally did not develop until the last 2 sec preceding the imperative stimulus (Hillyard, 1969b). McAdam (1966) requested subjects to interrupt a 200 Hz tone 1.5 sec after its onset, and informed the subjects about their performance. The CNV became augmented in amplitude during the first 20 trials, then became smaller. The actual time estimation performance had improved rapidly during the first 20 trials and then reached a plateau. Low (1969) employed a Sidman reinforcement situation to teach subjects to recognize a temporal interval and to press a button between the beginning of the thirteenth and the end of the fifteenth second in order to avoid a shock stimulus. The CNV associated with the button presses began to develop about 3 sec preceding the appropriate interval.

It seems clear that CNV amplitude increases in states that may be regarded as involving heightened attention and is reduced when attention is directed away from a forthcoming stimulus or when the subject's general state tends to reduce the capacity for attentive activity. Chapman (1969) has drawn attention to the fact that, in his evoked-response studies of attention, he obtained evidence of the slow potential changes which would reflect CNV and would be in accord with the foregoing conclusion.

One of the most interesting and provocative statements made about the relation between CNV and attention is Grey Walter's (1969) contention that all persons are capable of CNV, but that the main issue is the discovery of the circumstances under which CNV can be elicited. This point of view has particular significance for studies of psychopathological states and will be given further consideration later.

COMMENT

The relationship between evoked activity and phenomena related to attention obviously involves numerous problems of definition and unresolved complexities at both the physiological and psychological levels. However, the evidence reviewed here indicates that most of the recordable electrical activities of the brain are markedly influenced by psychological variables that can be subsumed under the broad topic of attention. The major exception appears to be the early or primary aspect of the sensory evoked response, which is apparently not very susceptible to psychological manipulation. This appears to agree with the idea that the early components reflect information transmission activities, whereas the later ones reflect information processing.

Setting aside the technical problems involved in distinguishing cerebral activity from extracerebral contaminants and the difficulties of establishing that the responses of subjects are actually in accord with experimental instructions, it would appear that the late components of evoked responses, including the P_3 wave, bear an intimate relation to the orientation of the subject toward his environment. Furthermore, the very slow potentials, such as the CNV and the "readiness" potential, reflect events associated with expectancy, motor preparedness, interest, and motivation. Thus, although the constraints of physiological recording procedures can never permit biological measurements to match the complexity of psychological situations, it appears possible to obtain objective, quantitative correlates of complex behavioral events from scalp recordings.

From the standpoint of psychiatric interest, psychophysiological correlates of the kind considered in this chapter seem to be of great importance as a means of objectifying covert and "private" mental activity. For example, if one assumed that schizophrenic psychoses are associated with abnormalities of attention, it would be logical to expect deviant late evoked-response activity, CNV and P_3 waves in schizophrenic patients. However, it should be noted that, important as it may be, confirming such expectations would not do more than confirm the already known fact that the patient's behavior is deviant and that the usual physiological correlates of his behavior are also deviant. In other words, such evidence shows that the psychophysiological relations that hold in normals, hold in psychopathological states. It leaves unanswered the question about the primacy of pathological physiology in the chain of events leading to disorder.

Chapter 7

EEG—Evoked-Response Relationships

There has been a tendency to disregard the EEG activity accompanying evoked responses, because the averaging process treats it as undesired "noise." The important relationships between evoked activity and the EEG in states of impaired consciousness indicate that such disregard is in error (Chapter 5). Furthermore, several writers have noted that the EEG phenomena associated with varying attentiveness in the waking state also may be intimately correlated with event-related potentials (Bergamini and Bergamasco, 1967; Näätänen, 1967; Tecce, 1970).

Although methods for automatic quantification of the EEG have a long history (Brazier, 1961), extensive application of computer techniques to EEG analysis seems to have taken place at about the same time as the development of evoked response averaging methods. The requirements for EEG quantification are, in some ways, more difficult and we are currently witnessing the application of the general-purpose computer to quantitative analysis of both EEG and evoked activities.

This chapter attempts to review some of the work relating evoked response and EEG characteristics. Relevant data have been obtained in two general ways. In one, evoked-response characteristics are correlated with those of the EEG taken separately, either in the same or in a different recording session. Studies employing this approach assume that the measurement of electrical activity will reflect some trait of the individual and that *between-subjects* correlations will indicate the degree of association between different electrophysiological characteristics. The other approach is to examine the relationship between averaged responses and the short EEG samples associated with each stimulus. For example, separate EEG samples may be computed for those trials in which EEG amplitude was above and below the subject's average during the recording session. The second approach tends to attribute variation in evoked activity to variation in the background EEG, and uses a *within-subject* approach to establish such relationships.

The material of the preceding two chapters generally supports the view that evoked-response characteristics vary as a function of state of alertness or awareness. This poses a problem if one wishes to study individual differences due to other factors. Since the EEG is often accepted as a technique for monitoring state of awareness, can it be used for this purpose during evoked-response recording? A positive answer to this question would allow one to use the EEG to equate subjects for degree of alertness. Obviously, since there are wide individual differences in the characteristics of EEG's taken during apparently similar states of alertness, the question cannot be answered simply. Although the value of EEG monitoring during drowsiness and sleep seems well established, the same cannot be said for its use to measure variations of alertness in the waking state.

BETWEEN-SUBJECTS CORRELATIONS

The most notable relationship between an evoked-response characteristic and background EEG activity occurs with respect to the after-rhythm of the visual response. This has been discussed in Chapter 3. In general, the after-rhythm is more prominent in the presence of regular and well-developed EEG alpha activity and its frequency is of the same order as that of the background EEG (Dustman and Beck, 1963).

Some unpublished observations from our laboratory provide additional statistical support for the presence of a significant association between EEG and visual after-rhythm amplitude. The amplitude of a 10 min EEG sample, taken with eyes closed in a "relaxed" awake state, was measured by means of a Drohocki-type integrator (Goldstein and Beck, 1965; Häseth et al., 1969). The EEG was recorded immediately after a lengthy evoked-response test session in 38 healthy college students. To reduce artifact, the taped EEG was passed through a band-pass filter with upper and lower frequency settings at 2 and 30 Hz before amplitude integration. The averaged responses to 300 flash stimuli, usually obtained about one hour before EEG recording, were measured with a computer program that determined the average deviation from the mean of arbitrarily selected epochs following the flash. The correlation between EEG amplitude and the average deviation from the evoked response mean in the epoch 200 to 800 msec after the flash gave a coefficient of 0.601 ($p < 0.01$).

The evidence concerning the relation between early evoked-response events and EEG characteristics is less clear. Cigánek (1961a) and Ebe et al. (1962) found no relationship between EEG and photic evoked-response characteristics. Kooi and Bagchi (1964a) also reported that the amplitudes of occipitoparietal and vertex "sharp" waves were unrelated to alpha frequency, alpha amplitude, or an alpha persistence index. However, in a more detailed analysis, they showed that two of five occipital response peaks measured during the initial 250 msec were of greater amplitude in those subjects with higher-voltage alpha in their resting records (Kooi and Bagchi, 1964*b*). The initial surface negative peak latency was found to be longer in subjects with low-voltage resting records and low alpha persistence during response summation. The latency of the third, occiput-negative peak was inversely related to resting alpha frequency. All significant correlations were of low order (0.23 to 0.27). The fact that significant correlations tend to be low may account for a number of additional reports of failure to detect a relationship between evoked response and background EEG characteristics (Tepas *et al.*, 1962; Chapman and Bragdon, 1964; Garcia-Austt, 1963; Werre and Smith, 1964).

One report does show strikingly positive findings relating EEG background activity to photic evoked responses (Rodin *et al.*, 1965). These workers employed a Faraday low-frequency analyzer to measure EEG amplitude for various frequency bands for a 40 sec period. They measured the evoked-response characteristics by means of a computer program that identified the number of positive and negative peaks, the time and amplitude of the earliest positive and negative peak, and the time and amplitude of maximum positive and negative peaking. Separate analyses were conducted for male and female subjects and for right and left occipital areas. The results confirm the generally higher evoked response amplitude found in female subjects. A large number of significant positive correlations between EEG and evoked-response amplitude were found. Although the magnitude of correlations was greater in females than in males, and in data from the right than from the left occipital area, the trend of the results was the same for both sexes and for both hemispheres. The maximum peak-to-peak evoked-response amplitude yielded correlations as high as 0.851 with the amount of energy in the various EEG frequency bands. A notable feature of the results was the demonstration that evoked-response amplitudes were correlated with the amount of energy in EEG frequency bands ranging from the slowest to the fastest, i.e., the correlations were not restricted to the alpha band. Another finding of interest is that the number of positive evoked-response peaks was correlated with the amount of energy in the fastest EEG frequency band.

The trend of our own results is in agreement with those of Rodin *et al.* since we found that the maximum peak-to-peak amplitude for the first 150 msec of the flash-evoked response yielded a correlation of 0.418 ($p < 0.01$) with "resting" EEG amplitude in a group of 40 normal subjects (Shagass *et al.*, 1968). We also found a significant correlation of 0.378 between somatosensory response amplitude (maximum peak-to-peak to 40 msec) and EEG amplitude. Cigánek (1965) has reported that auditory evoked-response amplitude is positively related to EEG amplitude and a trend in this direction has been noted in data obtained by Callaway (Shagass *et al.*, 1968).

Comment

The variable results obtained in attempts to relate EEG and evoked-response amplitude, particularly with respect to the first portion of the response, may to some extent be a consequence of the methods used for measuring the EEG, since most of the positive results have been obtained in studies employing automatic EEG quantification techniques. The nonautomatic methods employed in most of the negative studies involve difficulties of sampling and subjective error. The specifics of recording procedures have also varied from one study to another; this would contribute to discrepant findings.

The available data suggest that there is probably a genuine tendency for subjects with greater EEG amplitude to have higher evoked-response amplitude, although the magnitude of the correlation may be rather small. The correlation may also, to some extent, be artifactual in the sense that it may result from the fact that the electrical signals must pass through the same extracerebral tissues to be recorded.

EXPERIMENTAL MANIPULATION OF THE EEG

The simplest way to manipulate the EEG is to ask the subject to open or close his eyes. The visual after-rhythm is clearly affected by such maneuvers, and its incidence drops from about 80 % when eyes are closed to 20 % when the eyes are opened (Cohn, 1964). There is also some indication that the earlier visual response components are affected by such a maneuver with higher amplitudes being observed when the eyes closed (Domino and Corssen, 1964; Rodin et al., 1964). This observation could, however, reflect greater dilation of the pupils when the eyes are closed, allowing more light to strike the retina.

A particularly intriguing experimental manipulation of the EEG has been reported by Spilker et al. (1969). They employed a group of seven experimental subjects who had been trained to control their alpha rhythms by Kamiya (1968). These subjects could maintain control of their alpha rhythms without auditory feedback. Period analysis showed much higher modal EEG frequencies when they attempted to maintain low alpha than when they tried to keep alpha high. Spilker et al. found that all seven subjects had greater amplitudes in response to sine-wave light when there was high or abundant alpha in EEG than when the alpha was low. They also measured the evoked response to flash and found that two of the early waves of the flash response were greater in amplitude during periods of high alpha. The findings were unaffected by application of a cyclopegic agent or by varying the frequency of sine wave light stimulation. Spilker et al. also studied the auditory responses of three of their subjects and found that they did not vary significantly with changes from high to low alpha. The authors suggested that their findings relating alpha to visual evoked-response amplitude may reflect an influence of the state of attention on both electrophysiological indicators.

Since Spilker *et al.* failed to find auditory-response changes with alpha variation and since we found no somatosensory-response changes accompanying the marked reduction in EEG amplitude with TV viewing (Shagass *et al.*, 1970), it seems as though the visual response may be more related to EEG variations than those in other modalities.

WITHIN-SUBJECT RELATIONSHIPS

Levonian (1966) averaged EEG samples containing individual visual-evoked responses with respect to subsequent alpha frequency. He demonstrated that the averages differed considerably in relation to EEG frequency. Similar differences may be shown when EEG amplitude is made the criterion for sorting responses into averages. Figure 85 gives examples of evoked-response variations with both EEG frequency and amplitude. Näätänen (1967) measured EEG amplitude in relation to different states of attention; he found that there was reduced amplitude when the subject was in a state of general expectancy, and used his EEG observations to support the view that the enhanced amplitude of the late evoked-response components, associated with heightened attentiveness, reflects general arousal. Observations such as these suggest that the partitioning of evoked responses into separate averages, associated with differing amplitude and frequency characteristics of the contiguous EEG sample, might yield consistent relation-

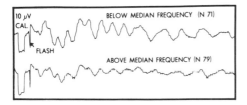

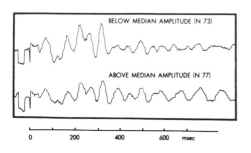

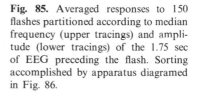

Fig. 85. Averaged responses to 150 flashes partitioned according to median frequency (upper tracings) and amplitude (lower tracings) of the 1.75 sec of EEG preceding the flash. Sorting accomplished by apparatus diagramed in Fig. 86.

ships. We conducted an analysis of the tape-recorded data gathered by Haseth *et al.* (1969) to examine this possibility.

The analysis utilized an electronic sorting device which is schematically described in Figure 86. In essence, the EEG sample occurring before each stimulus, and after the preceding evoked response was probably terminated, was first passed through a band-pass filter and then through either a baseline cross detector for frequency analysis or an absolute value generator for amplitude analysis. The sort criteria were set at the median of the baseline cross or absolute value counts obtained for the total sample. The EEG containing the evoked response was then placed into one or other channel of the averager depending upon the characteristics of the preceding sample. Responses evoked by median nerve stimulation and flash were studied. The EEG providing the sort criteria for the flash response was obtained from midline leads at the inion, and 6 and 12 cm above it. The EEG for sorting the somatosensory response was obtained from the leads at the inion and 6 cm above it, and from our usual bipolar somatosensory lead placement (Figure 13). The maximum peak-to-peak amplitude to 150 msec was determined for the responses to flash obtained from both pairs of midline electrodes. For the somatosensory response, the maximum in the first 40 msec was measured, and also the peak-to-peak amplitude of late activity after 100 msec.

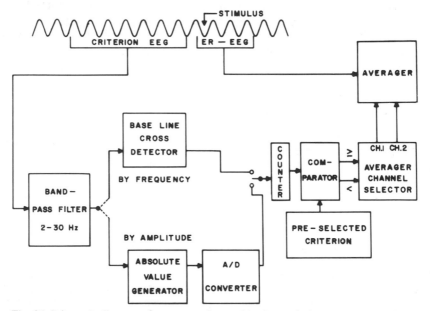

Fig. 86. Schematic diagram of apparatus for partitioning evoked responses according to preceding EEG. All of the EEG samples are first scanned in order to determine the best available approximation to the median frequency or amplitude. The preselected criterion is then employed to route the EEG samples containing the evoked response into the appropriate channel of the averager.

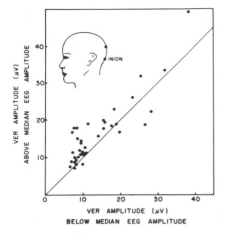

Fig. 87. Relationship between visually evoked response amplitudes associated with above and below median amplitude of preceding EEG samples. Evoked-response amplitude measured as peak-to-peak maximum for 150 msec after the flash. Both evoked responses and criterion EEG recorded from leads shown in inset diagram. Regression line drawn to indicate a perfect correlation. The preponderance of dots above the regression line shows that, in the majority of cases, evoked-response amplitude was greater when EEG amplitude was above median.

Although variations of the kind shown in Figure 85 were often observed the trend of the results was far from uniform. Figure 87 shows the most consistent finding, which indicates that flash evoked-response amplitude was greater when the amplitude of the preceding EEG sample was above average; this relation occurred in 30 of 37 subjects. There was also a tendency for the evoked response from the same electrode pair to be of greater amplitude when the preceding frequency was lower than when it was higher; this occurred in 25 of 37 subjects. However, these results were obtained only between the EEG and the evoked response from the electrode pair shown in Figure 87. When the EEG recorded from the adjacent midline pair of electrodes was used for the sort criterion, there were no consistent relationships between EEG characteristics and the amplitude of evoked responses recorded from either pair of midline leads. We also failed to obtain any significant relationships between somatosensory response amplitude and EEG characteristics.

Comment

These results indicate that the relationships between EEG and evoked-response characteristics are far from simple. Both sensory modality and electrode location appear to be important. Some relationships were found for the visual response, but none for the somatosensory response. Even for the visual response, a relationship found for a given pair of electrodes was not present when the EEG sort criterion was derived from an adjacent area with one electrode in common. Also, even when results were positive, there were several exceptions. The factors contributing to this variability of findings are not known. Personal characteristics of the subject and variable reactions to the experimental situation may be important. At any rate, the findings hardly provide strong support for the view that a large proportion of evoked-response variability in the waking subject can be explained by fluctuations in the associated EEG.

EEG–EVOKED-RESPONSE RELATIONSHIP AS A PSYCHOPHYSIOLOGICAL VARIABLE

The evidence concerning between-subjects correlations suggested the presence of a tendency for the amplitude of EEG and evoked-response events to covary, although the association was not easily demonstrable in some studies. Clearly, some subjects display either high or low amplitude in both their EEG and evoked responses, whereas others show no relationship. These marked individual differences suggested to us that the EEG–evoked-response amplitude relationship might, in itself, provide an electrophysiological variable of interest, and that it would be worthwhile to explore its possible psychological correlates. Relevant data were obtained in the study of Häseth *et al.* (1969) to which reference has already been made. In addition to the separately recorded EEG and evoked-response measurements, the subjects were given a battery of tests measuring perceptual and cognitive performance and some aspects of personality.

The tests that are relevant in the present context are those aimed at assessing perceptual functioning. They were as follows: (1) *Lifted weight discrimination* was tested using identical cylindrical jars. The standard weight was 100 grams and comparison weights were 90, 95, 100, 105, and 110 grams. To determine the threshold, the standard was compared with each comparison weight 20 times in random order and the best fitting straight line relating weights to category judgments was determined by the method of least squares. The probable error (*PE*) in grams was determined from this line (Woodworth and Schlosberg, 1954). (2) *Letter recognition* was measured by Thurstone's (1944) speed of perception test. Single letters, at varying angles from a central fixation point, were placed on 35 different slides and these were projected by means of a Kodak Carousel projector equipped with a photographic shutter. The exposure time was 0.02 sec. (3) *Line difference discrimination* was tested by asking the subject to compare a standard 88 mm-line drawn on a card with other lines that were variably shorter by amounts ranging from 1 to 20 %. (4) *Closure flexibility* and (5) *closure speed* were more complicated tests of visual–perceptual functioning. Closure flexibility (Thurstone and Jeffrey, 1965), or identification of concealed figures, is based upon the original Gottschaldt (1929) embedded figures. Closure speed (Thurstone and Jeffrey, 1966) presents a subject with incomplete figures which he must identify.

The matrix of linear correlations between the variables showed EEG amplitude to be correlated with both somatosensory and visual response amplitude (coefficients of 0.378 and 0.418, respectively). Also, EEG amplitude gave a correlation of -0.321 ($p < 0.05$) with the lifted weight discrimination measure; this indicated a trend for subjects with higher EEG amplitude to show better lifted weight performance. Individual evoked-response measures gave no significant correlations with indicators of perceptual performance.

To examine the possibility that EEG–evoked-response amplitude relationships differ with respect to perceptual performance, the correlations between EEG, somatosensory, and visual evoked-response amplitudes were determined separately for subjects with scores above and below the median on the letter recognition and lifted-weight discrimination tasks. Figures 88 and 89 show the data for the visual and somatosensory responses, respectively. For subjects with above median accuracy in perceiving letters, the correlation between the visual response and EEG amplitudes was 0.58 ($p < 0.01$), whereas the correlation for those with below average performance was not significant. Similarly, the visual response–EEG amplitude correlation was 0.60 for subjects with superior lifted-weight discrimination, while it was not significant for those with poor discrimination. The somatosensory–EEG correlation data were like those for the visual response; coefficients of 0.50 were obtained for correlations between evoked response and EEG amplitudes for subjects with superior perceptual performance, whereas the subjects with poor performance had nonsignificant correlations. For the above-average perceivers, multiple correlations between the amplitude of EEG and the two evoked responses were, respectively, 0.653, 0.693, and 0.717 for the letter recognition, lifted-weight discrimination, and line-difference discrimination tests. The corresponding correlations in the groups with below average performance were not significant.

The more complex perceptual tests yielded results that tended to be opposite in direction from those obtained with the simple perceptual performances. Thus the multiple correlation coefficients between the electrophysiological measures for subjects with below-median closure flexibility and closure speed were 0.683 and 0.718, respectively; the corresponding

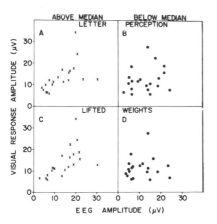

Fig. 88. Scatter plots relating EEG and visually evoked response amplitudes in subjects divided according to perceptual performance. A, subjects recognizing 16 or more letters; B, subjects recognizing 15 or fewer letters; C, subjects with lifted weight thresholds under 4.7 grams; D, subjects with thresholds of 4.7 grams or more. Note significant correlation between amplitudes when perceptual performance was above median. (Reprinted with permission from Shagass et al., 1968.)

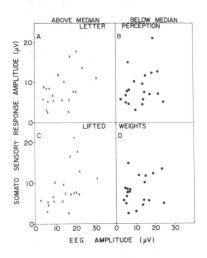

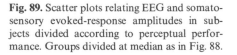

Fig. 89. Scatter plots relating EEG and somato-sensory evoked-response amplitudes in subjects divided according to perceptual performance. Groups divided at median as in Fig. 88.

values for the above-average groups were 0.446 and 0.393. Also the multiple correlation for those scoring below median on the Raven test of intelligence (Raven, 1958) was 0.732, while the coefficient was only 0.289 for those with above-median scores. Since the closure flexibility tests are significantly correlated with intellectual performance, this suggests that the EEG-evoked response concordance measures may bear different relationships to simple and more complex perceptual and cognitive functioning.

A major uncertainty in the results ensued from the fact that the various differences between correlation coefficients were not statistically significant. However, two sources yielded assurance that the results with the tests of simple perceptual functioning had not arisen by chance. One source was the data analysis giving the results shown in Figure 90. The subjects were classified into three categories of perceptual performance according to whether their scores were above the group median in both, one, or none of the lifted-weight and letter-recognition tests. Their electrophysiological results were classed as concordant when the amplitude values for the visual and somatosensory responses and EEG were all either above or below the group median. As Figure 90 shows, the majority of subjects who performed above average on both perceptual tests displayed electrophysiological concordance, whereas very few of the subjects who were below average on both tests did so. The relationship was significant at the 0.01 level.

Additional assurance that electrophysiological concordance is related to simple perceptual performance was obtained when the findings were confirmed elsewhere. Callaway and Jones, in conjunction with their two-tone auditory evoked-response studies, in a group of 25 schizophrenic patients, had also measured EEG amplitude and two-flash discrimination performance (Venables, 1963). Partitioning the group into two halves, on

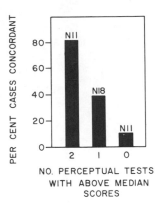

Fig. 90. Concordance between visual and somatosensory evoked-response and EEG amplitudes (all values either above or below median) related to performance on two perceptual tests. "Above median" test scores indicate a high number of correctly perceived letters and low lifted-weight threshold. (Reprinted with permission from Shagass *et al.,* 1968.)

the basis of above or below average two-flash performance, Callaway and Jones also found a significant correlation between auditory evoked response and EEG amplitude in the group with superior discrimination performance and a very low correlation in the group with poor two-flash performance (Shagass *et al.,* 1968).

Comment

In speculating about the significance of the EEG–evoked response concordance measure, it would seem reasonable to suppose that both variables contributing to it reflect activities that are under regulatory control of such central structures as the reticular formation. High concordance might then be taken as an index of the "tightness" of such regulatory control. The fact that concordance relationships were not sensory-modality-specific, for either the evoked response or the perceptual tests, seems to agree with the idea of "nonspecific" central regulatory mechanisms. The data suggest that such control is "tighter" in subjects with superior ability in simple perceptual tests. On the other hand, if true, the reverse tendencies for complex perceptual and cognitive tasks would be much more difficult to explain, particularly if one accepts the assumptions of the "tight" control formulation.

At present, the main significance of the EEG–evoked response concordance findings probably lies in the fact that they draw attention to the importance of looking at the relationships between electrical events as variables in themselves. The relationship may provide more significant psychophysiological indicators than the primary variables.

Chapter 8

Intelligence and Personality

INTELLIGENCE

Chalke and Ertl (1965) were the first to report evoked-potential correlates of intelligence. They compared the latencies of visually-evoked responses, recorded from bipolar leads placed over the left motor area, in three groups of subjects: 33 students with superior IQ's; 11 subjects with IQ's in the "low-average" range; and 4 mental retardates with IQ's ranging from 50 to 65. Three of the five latencies measured (peaks at 142 to 374 msec) yielded significant differences between the groups; the less intelligent subjects had longer latencies. Although their findings are of great interest, they are subject to criticism on several grounds. The retardate group was small. Age was poorly controlled and sex not stated. The averaging technique was somewhat unorthodox, as was lead placement. However, Chalke and Ertl noted that the latency measurements obtained with their technique agreed well with those obtained with a digital averager, and they selected their lead placement with the idea of "maximizing the input–output delay and thus to enhance the possibility of actually measuring central processing time."

In other studies, Ertl and his collaborators have obtained data confirming the conclusion that higher intelligence is associated with faster evoked response latencies. Barry and Ertl (1966) found correlations ranging from −0.76 to −0.88 between Wechsler IQ and latencies of the third and fourth visual response peaks in two samples of university students. Ertl and Schafer (1969) reported the results in a sample composed of 567 school children randomly selected from a population of 8000 attending grades 2, 3, 4, 5, 7, and 8 of the Ottawa school system. A digital averager was employed. Intelligence was tested with the Wechsler Scale for Children, the primary mental abilities test and the Otis mental ability test. The coefficients of correlation between latencies of visual response peaks averaging

77, 120, and 187 msec and the various measures of intelligence ranged from − 0.28 to − 0.35, and were all highly significant statistically. The puzzling feature of the reported results lies in the complete absence of correlation between the evoked-response latencies and age; some correlation would have been expected from the developmental data of other workers (Chapter 4).

Shucard (1969) has recently provided results generally consistent with those of Ertl. He studied over 100 subjects who were reasonably representative of the population at large. Visually-evoked responses were obtained under three conditions designed to introduce varying levels of "activation." The greatest activation was provided by a reaction time situation; for intermediate activation, the subject was required to count stimuli; the third condition was rest. Visual evoked-response amplitude and latency measurements were correlated with scores on a series of tests designed to reflect various aspects of mental ability. The results generally showed negative correlations between evoked-response latency and intellectual ability. Although significant, the correlations were small; the highest was −0.36. Higher correlations were obtained in the condition of low activation, which resembles more that employed by Ertl than in the others. The nature of the correlations obtained by Shucard suggested that they probably resulted from an association between evoked-response latency and such simple abilities as perceptual–clerical speed, span of apprehension, and visualization.

Whitaker et al. (1967) empirically derived a measure from the band-pass filtered photic evoked response that yielded high correlations with psychometric intelligence. The measure reflects the closeness of harmonic relationships between theta, alpha, and beta frequencies. A very high IQ would be predicted if the ratios of wave durations in these bands were exactly 4:2:1. Although the initial results reported by Whitaker et al. showed replication of their findings in several samples, more recent work has failed to confirm the original correlations (Whitaker, personal communication).

Following publication of Chalke and Ertl's (1965) results, we examined the relationship between evoked-response measurements and intelligence in those of our subjects for whom the information was available (Shagass, 1967b). We had Wechsler IQ estimates and visual-evoked responses for 19 psychiatric patients and 18 nonpatient subjects aged 65 to 80 years. Among the patients, the only significant finding was that the latency of the early negative wave peaking at about 30 msec was shorter in subjects with high IQ. More significant results were obtained for the nonpatients; correlations ranging from 0.46 to 0.75 were found for peaks 2 to 5. However, the direction of these correlations was opposite to that expected, since higher IQ was associated with longer latency. Only one significant correlation was found with visual-response amplitude. Also, we found no significant correlations for the somatosensory response in 17 nonpatient subjects. Our own results, therefore, do not agree with those of Ertl and

Shucard; in fact, where they are significant they are opposite in direction. However, our data came from rather restricted and specially selected groups that cannot be considered representative of the population at large.

Rhodes *et al.* (1969) compared visual responses in 10- and 11-year-old children with IQ ranging from 70 to 90 with those of children whose IQ ranged between 120 and 140. An occipital component with peak latency of about 250 msec occurred significantly early in the brighter than in the duller children, supporting Ertl's findings. Rhodes *et al.* also found that the components with latencies from 100 to 250 msec were larger for the bright children. Furthermore, there were greater differences between the amplitudes of responses, in the two hemispheres, in the bright than the dull children. In general, the right-central responses were consistently larger than those from the left-central area and this asymmetry was significantly greater in the bright children. The finding of lower amplitude in late components for the dull children is not consistent with the results in Down's syndrome, obtained in the same laboratory, since the mongoloids are reported as having higher amplitudes (Bigum *et al.*, 1970). This discrepancy may be important, if verified, since it suggests that evoked-response deviations in Down's syndrome may reflect processes other than those associated with lower intelligence.

Callaway and Jones (1970) have also reported relationships between evoked-response characteristics and cognitive functioning in children. Their evoked-response measures included both the two-tone auditory response and the slope of response amplitude to different depths of modulation of sine-wave light. They found strong correlations between age and all of their variables except response to sine-wave-modulated light. With age partialled out, there were a number of significant correlations between two-tone difference scores and mental test results; these generally showed better performance in subjects with less auditory response variability. Auditory response amplitude and latency gave few significant correlations. The sine-wave light response was also poorly correlated with the performance measures.

Mental Retardation

Barnet and Lodge (1967) found that click-evoked responses were significantly larger in infants with Down's syndrome than in matched controls. Most of the recordings were made with the infants in a sleeping state. Bigum *et al.* (1970) studied older mongoloid children. They found longer than normal latencies in the mongoloids, and also larger than normal amplitudes of late components in visual and somatosensory responses. The results in our laboratory with adolescent and young-adult mongoloids agree, in general, with those of Bigum *et al.* Amplitudes, particularly of late components, were larger for visual and auditory responses (Straumanis *et al.*, 1969*b*). Although we found no striking latency differences, we observed an interesting trend in relation to the latency of the initial somatosensory component. Whereas, in the normal population, this latency value is highly correlated with arm length (r about 0.8), the correlation was significantly lower in the patients with Down's syndrome. This suggests some factor interfering with the normal relationship between length of conduction pathway and cortical response latency.

From a psychometric point of view, the patients with severe brain syndrome studied by Straumanis *et al.* (1965) would be considered mentally retarded. It is of interest, therefore, that the visual responses in this group were distinguished from normal by prolonged latencies of late wave and marked reduction of afterrhythm.

Comment

The balance of evidence suggests that some association exists between evoked-response characteristics and cognitive ability. In some ways the intelligence studies are reminiscent of the history of investigations relating EEG characteristics to intelligence (Ellingson, 1966). Initially positive results have tended to be followed by less positive or even negative findings. However, since the data in the reasonably representative samples obtained by both Shucard (1969) and Ertl and Schafer (1969) confirm one another, there seems to be some firm basis for believing that there is a true relationship between psychometric intelligence and evoked-response latencies, even though its magnitude may be small. On the other hand, it is unfortunate that the positive results obtained so far have been based on visual responses recorded without first immobilizing the pupils. Since pupillary diameter appears to be related to cognitive functioning (Kahneman and Peavler, 1969), this leaves open the possibility that the correlations with IQ may depend upon a peripheral factor.

The results in mental retardation, particularly in Down's syndrome, suggest that evoked responses may reflect some brain events associated with severe cognitive impairment. Much more research in this area is needed; it will be of special importance to establish whether evoked responses in Down's syndrome are different from those in retardation syndromes involving the same IQ level without the specific etiology.

QUESTIONNAIRE TESTS OF PERSONALITY

The material in this section and the one that follows is derived mainly from two studies carried out in our laboratory. In one, responses to flash were measured in 19 nonpatients and 74 psychiatric patients falling into various diagnostic categories (Shagass *et al.,* 1965). In the other, somatosensory recovery functions were measured in 178 patients and 89 nonpatients (Shagass, 1968a, 1968b; Shagass and Canter, 1966). Questionnaire tests of personality and various other tests aimed at measuring aspects of personality functioning were administered. Not all tests were given to all subjects and, in final statistical analyses, the numbers were further reduced by the need to discard subjects to achieve proportionality in multifactorial analyses of variance in which age and sex were criterion variables in addition to test scores.

Maudsley Personality Inventory

The Maudsley Personality Inventory (MPI) (Eysenck, 1959) yields scores for extraversion (E) and neuroticism (N). Since psychiatric patient populations tend to have high N and low E scores, such scores in nonpatient subjects raise the suspicion of possible psychiatric difficulties and make the test useful for screening "controls."

The MPI was administered to a heterogeneous population of 46 patients and 12 nonpatient subjects in the study of visually evoked responses. Three amplitude and four latency measures were obtained. The latency of the first positive peak (about 45 msec) was significantly shorter in subjects with below-median E and above-median N scores. High-N-score subjects also had significantly greater amplitude of the initial negative–positive component. Since the patients had higher N and lower E scores than nonpatients, and since patients tended to have higher amplitudes and faster latencies than nonpatients, it seemed that the MPI data contributed little more than a confirmation of evoked response differences in relation to psychopathology.

MPI scores were correlated with somatosensory evoked-response variables in nonpatients only (Shagass and Schwartz, 1965b). The results of greatest interest emerged from an analysis of variance relating the E score to amplitude of the response to the unpaired stimulus (R_1). There were significant interactions between E score and age for the amplitudes of peaks 1, 4, 5, and 6 (Figure 24). The results are illustrated in Figure 91. In subjects aged 15 to 19, above median E was accompanied by high amplitude of the initial components, whereas the reverse was true in subjects aged 40 years or more. In subjects aged 20 to 39, there appeared to be no differences between high- and low-E subjects.

These results suggested the speculative possibility that the evoked responses could be reflecting a personality-related neurophysiological manifestation of aging. Since evoked response amplitude seems to maximal in childhood, minimal in the third and fourth decades of life, and greater after age 40, the relationship between amplitude and aging is U-shaped. The data in Figure 91 suggest that extraverted individuals may mature later and age later neurophysiologically than those who give introverted responses, i.e., their U-shaped curve would be shifted to the right of the introvert curve. However, a more recent study, confined to 17- to 19-year-old college students, failed to confirm the previously obtained differences

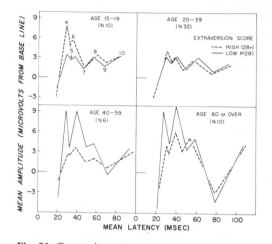

Fig. 91. Composite mean somatosensory responses for subjects in different age groups with extraversion *(E)* scores above and below median on the Maudsley Personality Inventory. Note that in ages 15 to 19, high *E* is associated with larger evoked responses, whereas, from age 40 on, high *E* is associated with smaller-amplitude evoked responses. (Reprinted with permission from Shagass and Schwartz, 1965*b*.)

with respect to *E* in the adolescent group (Häseth *et al.,* 1969). It is uncertain whether the difference in results represents a false finding in the small sample of ten subjects aged 15 to 19, in the earlier study, or is due to the fact that the newer sample consisted entirely of older adolescents, in whom the relationship might be more like that in the 20- to 39-year-old group. It still seems possible that there may be differential aging in relation to extraversion in subjects over age 40, but the evidence for differential neurophysiological maturation in the adolescent age group is shaky.

Several somatosensory recovery function differences were also found. Figure 92 shows those results that attained statistical significance. Subjects

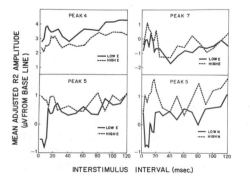

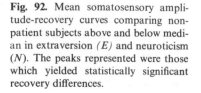

Fig. 92. Mean somatosensory amplitude-recovery curves comparing nonpatient subjects above and below median in extraversion *(E)* and neuroticism *(N)*. The peaks represented were those which yielded statistically significant recovery differences.

with above average E had less recovery of positive peak 4 for all portions of the curve. High-E subjects also showed greater positivity in one portion of the recovery curves for negative peaks 5 and 7. Although, theoretically, 5 and 7 are negative peaks, they tend to be seen above the zero line and, in patient-control comparisons, patients give lower values. These results suggest greater "excitability" in the more introverted subjects for the initial component, and possibly less for the later ones. Figure 92 also shows that peak-5 amplitude recovery values were more positive for subjects with high N scores. Since high N and low E tend to be associated, the results with the two MPI scales for peak 5 seem somewhat contradictory. The findings with the N scale are probably more trustworthy, since they held across the whole recovery curve and attained a higher level of statistical significance than those with the E scales.

Minnesota Multiphasic Personality Inventory

The Minnesota Multiphasic Personality Inventory (MMPI) was administered to 131 patients in the somatosensory recovery function study. The MMPI provides numerous scores, some of which are based on the original scales, derived by comparing nonpatients with clinical psychiatric diagnostic groups, and some of which have been derived in relation to other criteria. For our analyses, we selected the following MMPI measures: (a) distinction between psychosis and psychoneurosis using the rules proposed by Meehl and Dahlstrom (1960); (b) psychasthenia (*Pt*); (c) social introversion (*Si*); (d) ego strength (*Es*) (Barron, 1953); (e) modified *AR* novant (Welsh, 1965), which reflects to some extent the balance between impulsivity and dysphoria; (f) internalization ratio (*IR*), reflecting the balance between overcontrol and undercontrol; (g) R score, reflecting impulsivity; (h) A score, reflecting dysphoria and overcontrol (Dahlstrom and Welsh, 1960).

Considering the large number of analyses performed, relatively few significant relationships were found. Most of these were obtained with the recovery measures, but one difference in R_1 was of particular interest. The amplitudes of both the initial negative and positive components of the somatosensory R_1 were larger in patients rated as psychotic than those rated neurotic by the Meehl–Dahlstrom rules. This result was in accordance with previous predictions based on the notion that larger evoked responses reflect impaired inhibition. Since it has been suggested that superior perception is favored by inhibitory processes (Grundfest, 1959) and since perceptual activity is thought to be impaired in psychosis, it was predicted that psychotic patients would have larger evoked responses. Only one other R_1 amplitude difference was found; peak 8 differed with respect to ego strength, the group with the lowest third of Es scores having significantly smaller peak-8 amplitude. This finding is difficult to interpret.

The statistically significant recovery findings in relation to the MMPI criteria were as follows: (1) Subjects with good *Es* showed greater amplitude recovery for peaks 9 and 10, whereas those with intermediate *Es* scores showed greater recovery for peak 1. (2) Patients with a low internalization ratio gave less latency recovery for peak 10. (3) Subjects with high *A* score had greater latency recovery for peaks 5, 6, and 7 in the later segments of the recovery curve (70 to 120 msec). (4) Subjects with a high *R* score had more amplitude recovery on peak 1 and less latency recovery for peak 6. (5) Subjects with low *Pt* had more amplitude recovery on peaks 5, 6, and 10, and slower latency recovery on peak 1. (6) Patients with a low social introversion score had greater amplitude recovery for peak 10 and greater latency recovery for peak 9. The directional trends of these findings are summarized in Table III.

Since the number of significant relationships to MMPI variables exceeded that expected by chance alone, some of the findings probably indicate true associations. The genuine results would be more readily identifiable if they clustered in some recognizable pattern. Although no convincing pattern can be easily extracted, it is possible to derive the impression of some degree of consistency if we consider that high internalization ratio, *A, Pt,* and *Si* and low *R* scores should cluster together as indicators of introversion and dysphoria. In addition, we might expect low *Es* to be more common in such patients. From this point of view, our data show greater amplitude and slower latency recovery in patients who were less dysphoric, less introverted, and displayed better ego strength (Table III). Such patients would probably be more "normal" than the others, so that the trends obtained by dividing the patient group with respect to MMPI scores generally agree with those found in the patient–nonpatient comparisons to be presented in the following chapter.

Table III. Direction of Significant Recovery Findings in Relation to MMPI Scores in a Heterogeneous Psychiatric Population

MMPI score	Number of patients	Greater recovery[a]	
		Amplitude	Latency
IR	98		High
A	104		High
R	102	High	Low
Pt	110	Low	High
Si	102	Low	High
Es	104	High	

[a] Read as follows: "High" under latency for *IR* means that subjects with high *IR* scores had greater latency recovery.

PERCEPTUAL OR PERFORMANCE TESTS OF PERSONALITY

Bender Gestalt Performance

The Bender Gestalt test is used clinically in a number of ways. We scored the test results according to the method described by Pascal and Suttell (1951).

Figure 41 illustrates the visual-response results obtained in a sample of 11 nonpatients and 43 psychiatric patients of various kinds who received the Bender test. It also shows the differences between the total group of psychiatric patients and nonpatients. The trends for patient-control differences were more or less replicated in the comparisons between subjects with above- and below-average Bender performance. In general, the patients with poor Bender scores had higher amplitudes, that for the peak-to-peak measurement from 2 to 3 being significant. Latency 2 was significantly shorter in the subjects with poor Bender scores. These findings are similar to those obtained in subjects with low E and high N scores on the MPI. Again, because of the mixed nature of the subject population, and because Bender performance is generally poorer in a psychiatric population, it appeared that the results reflected a nonspecific association between evoked-response measures and presence of psychopathology.

Bender scores were available for 36 subjects in the somatosensory recovery study. Those with poor Bender performance had significantly more amplitude recovery for peak 4, a finding that seems inconsistent with the general trend for recovery to be reduced in psychiatric patients. However, a large portion of the subject group was composed of patients with severe chronic brain syndromes, in whom recovery of the initial component tended to be greater than normal. Additional positive findings with relation to Bender Gestalt performance were obtained only for the first 20 msec of the recovery curve; for this portion of the curve, poor performance was associated with increased recovery of peak 10 and reduced recovery of peaks 5 and 6.

It appears that our results with the Bender Gestalt test were heavily influenced by the fact that the populations studied were composed mainly of psychiatric patients. Since this visual–motor performance test seems to be sensitive to several aspects of psychopathology, the only conclusion that may be drawn is that evoked responses probably differ with psychopathology, but clues concerning the relevant psychological dimensions are not readily extracted from the data.

Critical Flicker Fusion (CFF)

CFF has been the subject of numerous investigations and has been related to many factors, including personality variables (Landis, 1954). We measured CFF threshold by means of a Krasno–Ivy flicker photometer. The average of three dial readings, in darkness, after 20 min dark adaptation and three measurements in light was taken as the threshold.

CFF thresholds in 92 psychiatric patients were correlated with somatosensory measurements. No significant relationships were found between CFF and R_1 measurements. However, CFF was found to be significantly related to several measures of recovery; these occurred mainly in the segment of the curve involving intervals from 70 to 120 msec. Subjects with low CFF showed more amplitude recovery for peaks 1 and 4, less amplitude recovery for peaks 5 and 6, and less latency recovery for peak 5. The amplitude-recovery findings seem to be similar to those obtained for Bender Gestalt scores. Poor Bender performance and low CFF threshold should probably go together, since both have been found in association with organic brain damage and very high levels of anxiety. However, as indicated previously, the findings for peaks 1 and 4 are not consistent with the patient–control differences to be described later. On the other hand, the findings for amplitudes 5 and 6 do agree with the results obtained in patient–control comparisons. For these amplitudes there was less recovery in the low-CFF group, which would be expected to be more pathological. It may be that both CFF and Bender Gestalt performance tap an "organicity" factor that is associated with greater amplitude recovery of the initial component.

Field Dependence–Independence

The rod-and-frame test described by Witkin *et al.* (1954) has been employed in many investigations of personality. The procedure is carried out in a dark room. A luminescent rod and frame, tilted at standardized angles from the vertical, is presented to the dark-adapted subject. The subject is required to judge the deviation of the rod from the vertical and instructs the experimenter to move the rod to the position he considers to be vertical. The score is the mean deviation of the judged from the true vertical. A high score is interpreted as field dependence, because the subject is considered to be influenced by the tilted frame, whereas a low score is interpreted as field independence.

We analyzed the relationships between rod-and-frame and visual-response measurements in 41 patients and 11 nonpatients. The only significant finding was an interaction between latency of peak 1 and flash intensity. Greater latency decrements with increasing intensity were associated with greater field independence. This finding is consistent with the results of a recent study by Buchsbaum and Silverman (1970), in which visual-response amplitudes were related to rod-and-frame test performance. The responses were evoked by tilted lines on a cathode-ray screen. Waveforms varied as a function of different degrees of tilt, but there were individual differences in the extent of such covariation. Subjects with low error scores on the rod-and-frame test (field-independent) showed the greatest visual response effect of different degrees of tilt; they also had less variability between trials. The differential effects were greatest in the portion of the response occurring from 250 to 500 msec after visual stimulus presentation.

Somatosensory results were analyzed in 92 patients. The most interesting finding was obtained with the initial component of R_1. Pursuing the same line of reasoning described earlier for the prediction that psychosis would be associated with larger evoked-response amplitudes, it had been predicted that field-dependent subjects should also have larger amplitudes. It was thought that high amplitude would signify less inhibitory activity, which might result in greater responsiveness to external cues in the field-dependent subject. The results did show significantly greater amplitude of the initial components in the more field-dependent subjects.

The recovery function findings showed greater amplitude recovery in the more field-dependent subjects for peaks 1, 4, and 10. These data also seem consistent with the formulation that field dependence may be associated with reduced inhibition.

Archimedes' Spiral

The duration of the afterimage induced by Archimedes' spiral was measured by means of the apparatus supplied by the Lafayette Instrument Company in 46 patients and 11 nonpatients. No visual evoked response characteristics were found to be significantly related to afterimage duration.

Stimulus Augmenting–Reducing

Several evoked-response studies have been carried out in relation to a personality variable based on concepts of stimulus intensity control. This variable has been called "augmenting–reducing." Petrie (1967) correlated "augmenting–reducing" with both pain threshold and ability to tolerate sensory deprivation. Stimulus reducers tolerate pain well, but tolerate sensory deprivation poorly; the converse is found for augmenters. Silverman (1967) has related the augmenter–reducer dimension to schizophrenia; he has emphasized the psychophysiological defensive function of stimulus reduction.

The behavioral test employed by Petrie to define augmenter–reducer tendency measures kinesthetic figural aftereffects (KFA). In one version of the KFA test, the blindfolded subject is seated in a chair with two long blocks of wood on the right and two on the left (Spilker and Callaway, 1969a). One of the blocks on the left is tapered, running from 0.5 inches anteriorly to 3.75 inches posteriorly; it is used for judging size. The subject slides his left-hand fingers in the holder along the taper of this block until he reaches the width which he feels to be the same as that of the 1.5-inch block between the fingers of his right hand. After a series of judgments of width,

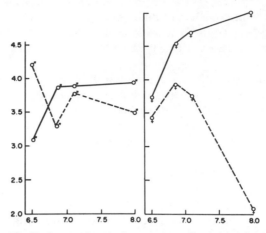

Fig. 93. Averaged evoked-response amplitude intensity functions for nonpsychiatric augmenter (solid line) and reducer (broken line) groups, separated by sex. Abscissa, log intensity in millilamberts; ordinate, amplitude of peak 4 in μv. (Reprinted with permission from Buchsbaum and Silverman, 1968.)

the subject rubs a 3-inch bar with his right hand and a 0.5-inch bar with his left hand back and forth to the beat of a metronome. After each 30 sec of rubbing, the subject makes a size estimate just as he did under prestimulation conditions. The score is the sum of the differences between estimated size before stimulation and after. A larger score indicates overestimation of size and is taken to reflect augmenting; a lower score represents reducing. Since the day-to-day variability of this estimation appears to be rather great in nonpatient subjects, there is reason to doubt that augmenting–reducing is an enduring cognitive style. However, the results obtained in relation to evoked-response measurements are of considerable interest.

Buchsbaum and Silverman (1968) correlated the results on the KFA test with measurements of visually-evoked responses to four stimulus intensities. Amplitudes were measured between two negative peaks occurring at about 75 and 200 msec latency, and a positive one at about 125 msec latency after the most intense flash. The latencies of the last two peaks were also used in data analysis. For the group as a whole, they found the usual relationships showing reduction of latency and increase of amplitude with increasing intensity of stimulus. However, dividing their nonpatient subjects with respect to scores on the KFA test, they found that the relationship between amplitude and intensity differed markedly in augmenters and reducers. The essential results are shown in Figure 93. The evoked-response difference between reducers and augmenters was much more marked for the female than the male nonpatient subjects. Female reducers actually showed a marked drop in response amplitude at the highest intensity of stimulation for the positive peak (peak 4) occurring at about 125 msec. Augmenters and reducers did not show important differences with respect to latency. The rank order correlation between the slope of the response amplitude–stimulus intensity function for peak 4 with KFA values was 0.63. Buchsbaum and Silverman also studied five male schizophrenic patients who were selected as extreme reducers. These subjects showed results very much like the female reducer subjects in Figure 93.

Blacker *et al.* (1968) employed a visual evoked response procedure comparable to that of Buchsbaum and Silverman to compare eight chronic LSD users with 16 nonpatient subjects. They found higher amplitude responses at the dimmer intensities in the LSD group, and also shorter latencies. However, the result of greatest interest was obtained when the normal group was divided into augmenters and reducers by means of the KFA test. The mean intensity–amplitude curves are shown in Figure 94. That for the LSD group was like that of the reducers. The correlation between the KFA score and the slope of these curves was about zero in

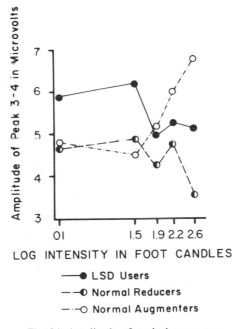

Fig. 94. Amplitude of evoked responses at five stimulus intensities for chronic LSD users and nonpatient subjects. Nonpatients divided according to their classification on kinesthetic task as reducers or augmenters. (Reprinted with permission from Blacker *et al.*, 1968.)

the LSD users, in contrast to a correlation of 0.73 in the nonpatients. The low correlation in the LSD group could be due to the uniformity of their evoked-response slopes. Blacker *et al.* found no difference between LSD users and normals on auditory evoked-response amplitude, latency, and two-tone similarity measures.

Spilker and Callaway (1969*a*) correlated the response to varying depths of sine-wave modulation of light with KFA performance. Figure 44 shows the kind of responses that they obtained. They computed a slope value reflecting the variation in visual response amplitude with different degrees of modulation. Rank correlation between this slope value and the KFA score was 0.46 for the response recorded from the left occipital area, 0.74 for that recorded from the right side, and 0.66 for the average for the two sides. Contrary to Petrie's theory that reducers should be sensitive to low intensities, Spilker and Callaway's data showed that the subjects with high KFA scores (augmenters) had greater response amplitudes at all intensities of stimulation used.

Although the test–retest correlation of both the KFA scores and the visual response slopes was rather low, Spilker and Callaway's data showed that in 9 of 11 subjects retested, both of these scores changed together. Some of the subjects were also tested with their pupils dilated by means of phenylephrine. In all subjects the visual response slope was greater after administration of the mydriatic. The most striking effect of dilating the pupils was a greater response amplitude at the greatest depth of modulation. This indicated that pupillary response was partially responsible for the reducing effect. However, in some subjects, the negative slope present before dilatation of the pupils still persisted after dilatation.

Level of "arousal" was among the factors considered possibly related to the variability of the response to sine-wave-modulated light from one test occasion to another by Spilker and Callaway. Consequently, they decided to experimentally manipulate level of "arousal" by administering CNS depressant and stimulant drugs (Spilker and Callaway, 1969*b*). They employed sodium pentobarbital and ethyl alcohol as depressants and methamphetamine and caffeine as stimulants. Their results demonstrated a significant reduction of the evoked-response-slope values after both depressants, but no effects of the stimulants. The authors concluded that their results supported the notion that "arousal" is a factor related to the "augmenting–reducing" phenomenon, but not the only one. Since depressants not only reduced the slope but increased the amplitude of the response to the low depths of modulation, the results appear to be consistent with the observations of Petrie (1967). The absence of effect of the stimulants was not satisfactorily explained.

The findings on augmenting–reducing seem to be of considerable importance because they suggest that evoked-response methods may be used to provide objective measurements of characteristic CNS modes of handling sensory intake. The fact that reasonable correlations have been demonstrated between visual evoked-response measurements and behavioral tests of kinesthetic response indicates that the relevant perceptual dimension is not limited to one sensory modality. However, from the standpoint of a useful personality measure, the high degree of variability in both the KFA and evoked-response measures is somewhat disconcerting. This variability has been shown mainly in nonpatient subjects. It could still be the case that, in association with a psychopathological state of relatively long duration, the tendency toward reducing or augmenting would be relatively fixed. Although our own results with different flash intensities failed to show any impressive slope differences between psychiatric patients and nonpatients, or between different types of patients, the flash that was employed was quite bright even at it lowest intensity (Shagass et al., 1965). Furthermore, we made no assessment of augmenting–reducing.

One serious methodological criticism of Buchsbaum and Silverman's work is that they did not immobilize the pupils of the subject. Since Spilker and Callaway (1969) did find some pupillary effect employing sine-wave-modulated light, which keeps mean brightness constant, the possibility that some of the augmenting–reducing differences in visual responses found by Buchsbaum and Silverman may reflect a peripheral pupillary effect cannot be ruled out.

"Repressiveness"

Shevrin and Fritzler (1968) rated "repressiveness" from Rorschach test performance and selected five pairs of twins who differed markedly in such ratings and five pairs whose ratings were identical. The tactile evoked-responses of these twins, which had been studied by Shevrin and Rennick (1967) in an experiment devoted to selective attention, were compared. Amplitudes of the twins high in "repressiveness" were greater for components peaking at averages of 113 and 207 msec than those of twins lower in "repressiveness," while the reverse was true for a later component with a mean latency of 524 msec.

More recently, Shevrin et al. (1969) employed the subliminal stimulation method of Shevrin and Fritzler (1968) (Chapter 6) to study the correlation between clinical and Rorschach indices of "repressiveness" and evoked response discrimination of subliminal stimuli. The component peaking positively at about 160 msec was negatively correlated with "repressiveness"; this correlation was about the same for the pen-knee figure as for the abstract drawing. However, the correlation was obtained only for the first 1 msec exposure sequence and not for a second (Figure 95). Also, there was a trend for correlations in the reverse direction with a supraliminal 30 msec exposure. An additional finding was that the amount of alpha activity present in the average response was negatively correlated with "repressiveness."

IMSEC(1)   30MSEC   IMSEC (2)

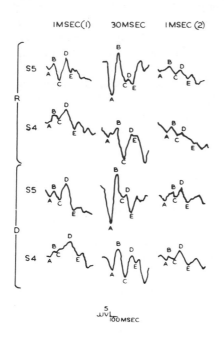

Fig. 95. Evoked responses in most and least repressive subjects. Subject 4 (*S4*) was the most and *S5* was the least repressive *S* in the sample. The upper pairs of tracings for the three exposure conditions are responses to stimulus *R* (meaningful, clearly defined). The lower pairs of curves are responses to the abstract stimulus *(D)*. (Reprinted with permission from Shevrin *et al.*, 1969.)

    The findings of Shevrin and his coworkers are of considerable interest, because they suggest that an important aspect of personality functioning, namely, the tendency to repress, may come into play virtually immediately upon presentation of sensory stimuli. There is some apparent inconsistency in the findings, since repressers had *smaller* amplitudes with subliminal stimuli and larger ones with suprathreshold stimuli. Shevrin *et al.* suggest that the repressive subject may be attending intensely to a supraliminal stimulus, while his scanning is diminished; with a subliminal stimulus, he may be attending to the rest of the field and discriminating it, although his overall responsiveness to subliminal stimuli is relatively low. Convincing explanations undoubtedly attend further studies.

    There may be some problems in interpretation of Shevrin's findings with subliminal stimuli. Since the pupil was not immobilized, the possibility that the evoked response differences could be peripherally produced cannot be ruled out. Secondly, the number of subjects was quite small and a significance level of 0.10 was accepted for some of the important correlations. Third, the difference between high- and low-repressive subjects seemed to be greatly reduced in a second 1 msec exposure sequence; this would suggest that the influence of "repressiveness" as a personality trend was exerted mainly during initial exposure to subliminal stimulation and that such influence was relatively transitory in nature. Although habituation may account for the changes in the second stimulus sequence, as suggested by the authors, this may involve a pupillary effect. Even if habituation occurred at a cerebral level, it would be necessary to relate evoked-response habituation to a personality trend that is presumably relatively enduring.

Although open to criticism, the experiments with subliminal stimulation and "repressiveness" provide an exceptionally interesting approach to evoked-response investigation of personality variables.

CONTINGENT NEGATIVE VARIATION

McCallum and Walter (1968) have reported significant negative correlations between CNV amplitude and Middlesex Hospital Questionnaire scores of anxiety, obsessionalism, and depression. The lower the amplitude during distraction, the higher the score on the psychopathological variable. Their results seem to be in conflict with those of Bostem *et al.* (1967) who found that questionnaire test measures of general psychopathology and obsessional anxiety, including the Cattell and MMPI, were unrelated to CNV morphology. However, Bostem *et al.* did not employ a distraction condition, which reduced CNV amplitude much more in patients than in nonpatients (McCallum and Walter, 1968).

Knott and Irwin (1968) selected groups scoring in the extreme range of the Taylor Manifest Anxiety Scale, which is composed of items selected from the MMPI. They found no difference in CNV amplitude between the high- and low-anxiety groups under circumstances where a motor response was required, where one was not required, and where the imperative stimulus was a low-intensity shock. The one difference between the groups occurred when the imperative stimulus was a high-intensity shock. In that condition, CNV amplitude was lower for high-anxiety subjects. Knott and Irwin interpreted this finding as resulting from "higher base-line cortical negativity" in the high-anxiety subjects which prevented them from producing additional negativity under stressful conditions. An alternative explanation would be that high-anxiety subjects would be less able to direct attention to the stimulus under conditions of stress and that amplitude would be lowered as it is under conditions of distraction.

The available evidence suggests that CNV may be diminished in amplitude under conditions producing stress or anxiety in subjects prone to such reactions.

SEDATION THRESHOLD

Sedation threshold (Shagass, 1954) is the name given to a test of reactivity to sodium amobarbital, which measures the amount of the drug required to reach a given level of sedative effect. The threshold criteria employed have been changes in the EEG, production of dysarthric speech, nystagmus, inability to respond to verbal stimulation, and inability to double digits (Shagass, 1967c; Claridge and Herrington, 1960). We studied the relationship between visual evoked-response characteristics and sedation threshold measured by the technique of Claridge and Herrington in 42 psychiatric patients (Shagass *et al.,* 1965). No significant differences were found between patients with high and low sedation thresholds, with one exception. The exception was in the form of a significant interaction for the latency of peak 2 which showed a greater decrement with higher intensity in subjects with low sedation thresholds. If greater latency decrement were taken as an indicator of an augmenting tendency, low sedation threshold, which is characteristic of extraverts and hysterics, would go with augmenting. However, we have no additional data to support this possibility.

COMMENT

It will be apparent that, although evoked-response characteristics have been found to be significantly related to various indicators of personality, most of the results are not sufficiently consistent or patterned to provide definite conclusions. One possible exception may be the data relating intensity-response functions to the augmenter–reducer dimension of sensory control. The results with the augmenter–reducer concept involve correlations of higher magnitude than those found with most other criteria and there has been some replication. Furthermore, the personality variable involved is more closely related conceptually to the neurophysiological measurements than most of the other variables that have been studied. On the other hand, the available evidence suggests that augmenting–reducing may be a transitory rather than an enduring characteristic, so that it may be determined more by the current situation and the current affect state of the individual than by his personality make-up.

The available results with personality tests are heavily influenced by the composition of the subject populations studied. Populations composed of patients with varying psychopathology will give results dependent upon the way in which the psychopathology influences the test findings. Apart from augmenting–reducing and "repressiveness," nonpatient subjects have been studied mainly with the Maudsley Personality Inventory (MPI). Here the result indicating that extraversion, at least in subjects over 40, is associated with smaller somatosensory evoked-response amplitude, is of considerable theoretical interest. If evoked-response amplitude is taken to reflect the balance between excitatory and inhibitory processes, such balance seems to be related to a complex behavioral pattern. The MPI *E* scale consists mainly of two kinds of questions; responses indicating high sociability and lack of undue concern about adequacy of performance give high *E* scores. If the data that we obtained are to be taken at face value, it would appear that persons over 40 who are more inclined to be in the company of others than alone, who enjoy social situations, and are less worried about their work function, tend to have smaller evoked responses and a greater degree of inhibitory neural functioning. The finding that they have less recovery of the initial somatosensory component would also be in accord with the notion of greater inhibitory activity. Also, the recovery curves for peak 4 in younger subjects are very like those of high-E subjects (compare Figures 55 and 92). The results in high-E subjects over age 40 seem to be in accord with the formulation that aging, as reflected in evoked-response indicators, may take place more slowly in this group. That the neurophysiology may parallel behavior seems plausible, since older people are generally regarded as being "younger" when they are more sociable and less fussy. The results with the Bender Gestalt test and CFF threshold determination agree to some extent with those obtained with the MPI in that they suggest that greater "organicity," which could signify earlier neurophysiological aging, is associated with greater recovery of the initial somatosensory component.

The two significant findings obtained with respect to the initial somato-sensory component in responses to unpaired stimuli involved higher amplitude in the field-dependent subjects and in patients rated as more psychotic by means of the MMPI. These findings were noteworthy mainly because they were predicted on the basis of ideas concerning the balance of excitatory and inhibitory processes reflected in evoked-response am-plitude. Both findings support the idea that reduced inhibitory activity is associated with less accurate perception or impaired responsiveness to environmental cues. Shevrin's results showing higher-amplitude responses to supraliminal stimuli in subjects high in "repressiveness" may also be in accord with this view, although they are based on later components.

The data relating visual evoked responses to Bender Gestalt and MPI scores and somatosensory recovery functions to several MMPI variables seem to fit into a general pattern involving the presence and severity of psychopathology. Greater amplitude and faster latency of visual evoked responses and reduced-amplitude recovery combined with faster latency recovery of somatosensory responses were all associated with evidence of greater introversion, greater dysphoria, and reduced ego strength. These trends are in the same direction as those found by comparing groups of patients and nonpatients. Since the subject populations involved were composed mainly of patients, it seems more parsimonious to interpret the test findings as correlates of the severity of psychopathology than to postulate specific personality trends independent of the presence of psycho-pathological states. However, should the same results emerge in nonpatient subject groups, the possibility that there are true evoked-response correlates of personality dimensions would receive support from the data so far gathered. As of now, the evidence showing that test scores indicating greater disorganization, more emotional distress, or more introversion in patients were associated with physiological findings more like those of the patients in patient–control comparisons, do little more than provide a more objective body of evidence reinforcing the findings based upon clinical categorization alone.

Future advances will depend to a considerable extent upon the availability of psychological techniques for personality measurement that lend themselves to clear interpretation. Even with their drawbacks, the augmenter–reducer results are encouraging as an example of what may be possible when a personality variable can be defined and measured with some degree of assurance.

Functional Psychiatric Disorders

The first application of evoked-response methods to the study of psychiatric patients was probably carried out in our laboratory (Shagass and Schwartz, 1961*b*). Although psychiatric diagnostic criteria present difficult problems of reliability, they offer convenient reference points that enable clinicians to relate laboratory findings to states with which they are familiar. Our research strategy has been to compare evoked-response characteristics in psychiatric diagnostic groups with the expectation that repeatable neurophysiological differences would provide a first step toward defining psychopathological dimensions based on aberrant physiology. In other words, having found a diagnostic difference, one may then attempt to determine which factors involved in the diagnosis account for the correlation with the electrophysiology.

EVOKED RESPONSES TO UNPAIRED STIMULI

Somatosensory

We began our earliest studies of psychiatric patients without adequate facilities for amplitude calibration, because they did not seem to be essential for the comparisons of relative amplitudes in recovery-function measurements. When amplitude calibration became available, we observed that large amplitude potentials were evoked by the first stimulus of the pair less often in nonpatients and in patients with psychoneurotic disorders characterized by anxiety, depression, and somatic complaints (dysthymics) than in the remaining patients. An inverse relationship was also found between mean R_1 amplitude and peak recovery of the initial component during the first 20 msec of the curve (Shagass and Schwartz, 1962*d*). These observations led to a systematic study in which the relationship between stimulus intensity and somatosensory-response amplitude was compared in various clinical groups. The initial report (Shagass and Schwartz, 1963*a*) dealt with findings in 24 nonpatients and 42 patients. In a subsequent paper (Shagass and Schwartz, 1963*b*), the size of the patient group was extended to 87.

Figure 96 shows the mean intensity–response curves for the initial negative–positive component (1–4, Figure 24) obtained for 24 nonpatients, 11 dysthymic patients, and 31 patients of other kinds. Whereas the curves were similar for nonpatients and dysthymics, those for the nondysthymic patients showed higher mean amplitude at every stimulus intensity. Figure 97 shows the mean curves obtained by plotting the data of Figure 96 with the logarithm of stimulus intensity as the abscissa. This transformation yielded approximately rectilinear curves, so that it was meaningful to compute the slope of the best fitting straight line for each subject. Median slopes were 4.2, 3.5, and 8.5 for nonpatients, dysthymics, and nondysthymic patients, respectively, and the nondysthymic group differed from the other two at a high level of statistical significance.

Using the response evoked by a stimulus about three times sensory threshold, the subject groups were also compared with respect to median amplitude of several later components as well as the primary. Significant differences were found for the primary for the first positive component after the primary and for a negative component peaking at an average of 123 msec after the stimulus. In all instances, the amplitude differences between nonpatients and dysthymics were not significant and both groups differed from the nondysthymic patients.

In the second report (Shagass and Schwartz, 1963*b*), the larger number of patients permitted further breakdown of the groups. The results showed steeper slopes and larger evoked responses for a given stimulus intensity in patients with diagnoses of conversion or dissociative reaction, situational or adjustment reaction, various personality disorders, schizophrenia, and psychotic depression. A small group of nine patients of diverse types, classified as miscellaneous, did not differ from the nonpatients or

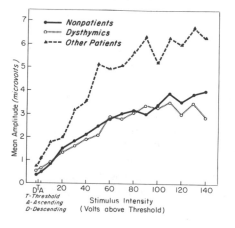

Fig. 96. Mean amplitude of initial negative–positive component of somatosensory evoked response as a function of stimulus intensity in 24 nonpatients, 11 dysthymic patients, and 31 patients of other kinds. (Reprinted with permission from Shagass and Schwartz, 1963*a*.)

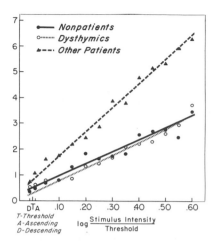

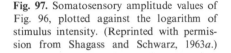

Fig. 97. Somatosensory amplitude values of Fig. 96, plotted against the logarithm of stimulus intensity. (Reprinted with permission from Shagass and Schwarz, 1963a.)

dysthymics. These results suggested that the intensity–response function essentially distinguished only two groups of subjects: (1) nonpatients and dysthymics; (2) all other patients. The slope of the intensity response gradient was not significantly correlated with age, nor did it differ significantly between the sexes. The reliability of the intensity–response measures was examined in 15 patients; the rank order correlation between two slope values obtained on different days yielded a coefficient of 0.60, while that for the mean amplitude of the curves was 0.82.

A subsequent study of somatosensory recovery functions, carried out on 178 psychiatric patients of various kinds and 89 nonpatient controls, presented an opportunity to check the amplitude findings for one intensity of stimulation (Shagass, 1968a). The response (R_1) to an unpaired stimulus 10 ma above sensory threshold, obtained during the recovery function measurement, was compared in various diagnostic groups. Eight latencies and amplitudes were measured, the latter as deviations from the estimated isoelectric line (peaks 1 and 4 to 10, Figure 24). The peak-to-peak amplitude corresponding to the initial component measured in previous studies, i.e., peak 1–4, was also obtained. The results for the R_1 measurements were remarkably negative. No significant amplitude or latency differences were found between patients and nonpatients. Furthermore, no significant R_1 differences were found in comparisons of matched groups of nonpatients, psychotics, and nonpsychotics, and in a further breakdown which compared nonpatients, psychoneurotics, and personality disorders matched for age and sex.

A few significant differences were found in analyses in which the patients in each of eight diagnostic groups were compared with nonpatients matched for age and sex. Mean peak 1–4 amplitude was greater in 13 patients with diagnoses of passive–aggressive personality trait disturbance, and the latency of peak 10 was greater in this patient group. Patients classified as dysthymic psychoneurotics had a longer latency for peak 7

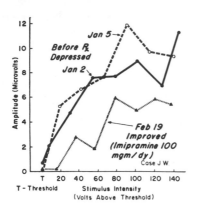

Fig. 98. Amplitude of initial negative–positive somatosensory component as a function of stimulus intensity in a patient with psychotic depressive reaction tested on three occasions. Note reduction of amplitude when patient was clinically improved after receiving imipramine for several weeks. (Reprinted with permission from Shagass *et al.*, 1962.)

than the nonpatients matched to them. Schizophrenic patients had a larger amplitude of peak 10 and a longer latency for peak 5 than their matched nonpatient controls. Considering the large number of statistical comparisons that were made, the significant findings were very few and one is disinclined to attach much weight to them.

Since the amplitude findings in the later study appear to be seriously discrepant with those obtained in the earlier intensity-response determinations, it seems important to consider possible reasons. A number of differences in method may be relevant. (1) The later R_1 amplitude measurements were obtained in the context of a recovery function study. (2) A full range of stimulus intensities was used in the earlier studies, whereas only a single intensity was used in the later one. (3) A constant-current stimulator was used in the later study, whereas the earlier ones employed a Grass S4 stimulator, which could give quite variable voltage drops across the subject, depending upon changes in his impedance. (4) Age and sex were controlled in the later study, but were only checked for possible statistical relationships in the earlier ones. As concerns the possible stimulator effect, it is relevant that the earlier intensity-response results were not related to variations in sensory threshold or in the minimal intensity required to elicit muscular twitch (Shagass and Schwartz, 1963*a*). In connection with the age factor, it could be supposed that unsuspected interactions between evoked-response amplitude and age may have given rise to some spurious findings. However, a special check of the 26 personality disorder patients, whose age was almost identical to that of the 24 nonpatients in the earlier study, revealed significant mean amplitude and slope differences; this could not have been due to inadequate control for age.

Although the later findings certainly cast doubt upon the validity of the earlier ones, additional data seem necessary to establish the true state of affairs. A new study should involve a full range of stimulus intensities, avoid the complications of paired stimuli, and employ a constant-current stimulus generator and an adequate number of subjects in different diagnostic groups matched for age and sex.

Additional evidence that suggests the merit of further study of the intensity–response function comes from a few serial tests carried out in patients before and after treatment. Figure 98 shows the results in one patient with psychotic depression who was treated with imipramine. The two curves (peak 1–4) determined prior to treatment were quite similar and of considerably higher amplitude than the one obtained when the patient was showing clinical improvement. Figure 99 shows three determinations in a patient with paranoid and depressive symptoms who was treated with electroconvulsive therapy (ECT). The measurements made between the fifth and sixth ECT, when the patient was improved but not fully remitted, yielded a curve of lower amplitude than that obtained before treatment; the curve obtained when the patient was in remission had the lowest amplitude of the three. These serial determinations are by no means conclusive, but they suggest that mean level and the slope of the intensity-response curve decrease with effective therapy and they are in accord with the group differences showing larger amplitudes in psychotic patients.

Heninger (1969*b*) measured somatosensory responses in a factorial study of treatment effects in depressive syndromes. He found that the amplitude of a component designated by him as 7–8, which probably corresponds to our 9–10, was significantly reduced after clinical improvement. Also, the latency of peak 7 was increased. The change in clinical rating of a factor identified as "withdrawn–retarded–depression" gave a correlation coefficient of 0.64 with the decrease in peak 7–8 amplitude. Heninger expressed the view that, although this late somatosensory component was in some way related to the manifestations of depression, the relationship was probably not specific.

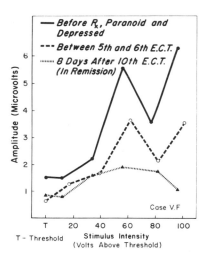

Fig. 99. Serial somatosensory intensity–response curves in a psychotic depressive patient treated with ECT. Note progressive reduction of amplitude with clinical improvement. (Reprinted with permission from Shagass *et al.*, 1962.)

Auditory Responses

The auditory response has not been a favorite in psychiatric studies. Available data have been reported in the context of relatively complex procedures, such as Callaway's two-tone test, or measurement of recovery functions. Satterfield's (1969b) results with recovery functions, comparing 22 depressed patients with 22 matched nonpatients, do permit comparison of amplitudes. Measuring between N_1 and P_2 of the click-evoked response, Satterfield found no significant amplitude difference, although there was a trend for the amplitudes of the patients to be greater than those of the controls, particularly at slow rates of stimulation. The groups also did not differ in the latencies of N_1 or P_2. Jones and Callaway (1970) found auditory response amplitudes, measured from P_2 to N_2, to be lower in schizophrenic patients than in nonpatients when a single tone was used, but not when two tones were used.

Visual Responses

In our study of visual responses, we recorded three average responses to unpaired flashes, applied to closed lids, at three different intensities (Shagass et al., 1965). However, even the weakest flash was quite bright. In the same experimental session, responses to unpaired flashes of light were also obtained in conjunction with recovery function determinations (Shagass and Schwartz, 1965a). Pupils were not immobilized.

Figure 41 compares the total group of patients with the nonpatients with respect to mean amplitudes and latencies of responses to single flashes at three intensities. In general, all three peak-to-peak amplitude measurements were greater and all latencies were less for the patients. The differences for amplitudes 2–3 and 3–4 were statistically significant. Mean R_1 amplitude 2–3, obtained in recovery-function determinations, which involved the intermediate intensity flash, was also significantly greater in patients. No latency differences achieved statistical significance. Breaking the patient group into psychotic and nonpsychotic categories, the nonpsychotics had greater amplitudes 1–2 and 2–3, and significantly longer peak 3 latency than the nonpatients for the R_1 recovery function measurements. The intensity–response measurements yielded the same differences between nonpsychotics and nonpatients and, in addition, the 3–4 amplitude measurement was significantly greater in the patients. No significant differences were found between the psychotic group as a whole and the nonpatients. In an additional analysis, the patient group was divided into: psychoneurosis, personality disorder, schizophrenia, and other psychosis. The R_1 values obtained during recovery measurement revealed a significantly shorter peak 2 latency in the schizophrenics than in the nonpatients, and showed that the high amplitudes found for the nonpsychotic group as a whole were about equal in psychoneuroses and personality disorders.

In addition to the foregoing measurements of the earlier visual-response components, we attempted to quantify the amount of after-rhythm occurring between 450 and 900 msec after the flash. The sample was started at 450 msec to avoid uncertainty concerning the time of onset of the after-rhythm. The number of oscillations which exceeded 2 μv in amplitude and had a duration of 60 to 140 msec was counted for each subject. Records with the requisite 1 sec analysis times were available for 60 patients and 16 nonpatients. The mean number of after-rhythm waves meeting the criterion was 3.94 in nonpatients and 2.38 in patients ($p < 0.01$). Within the patient population, the means of the psychoneurotic and schizophrenic groups, which were 2.54 and 1.80, respectively, differed significantly from those of nonpatients. It thus appeared that, although schizophrenic patients differed relatively little from nonpatients in the early portions of the evoked response to flash, they had much less after-rhythm. Since Straumanis et al. (1965) also found reduced after-rhythm in patients with severe brain syndromes, the possibility was suggested that this finding might be taken as evidence of some altered state of awareness in the schizophrenic patients. However, this interpretation does not seem to be in accord with the clinical findings of reasonably intact sensorium in the schizophrenics. Furthermore, after-rhythm diminution appears to accompany both impaired consciousness and heightened alertness. It seems more likely that the after-rhythm data may indicate reduced alertness in the brain syndromes and heightened alertness in the schizophrenics.

Visual-response characteristics have been studied in psychiatric patients by several groups of workers. Speck et al. (1966) found no significant differences between nonpatients and a psychiatric patient group composed mainly of schizophrenics in amplitude or latency of negative peak 3. There was, however, some trend for longer latencies in a small group of depressive patients. These workers also noted considerable interindividual variability in the evoked-response patterns of schizophrenic patients. They described three major patterns as follows: (1) "amorphous type" with low amplitude and broad deflections; (2) a group resembling the "average normal"; (3) a group with very high-voltage oscillating deflections. Floris and his collaborators (Floris et al., 1967, 1968) studied visual evoked responses and their recovery functions in normals, chronic paranoid schizophrenics, and neurotic depressives. They report no significant latency or amplitude differences between these two groups in response to unpaired flash. Their mean graph does show some tendency for earlier occurrence of peaks IV and V (Cigánek's classification) in the schizophrenics than in the other two groups. There is also some suggestion of lower amplitude between peaks V and VI in the neurotic depressive group. Ivanitsky and Arzuelova (1968) compared visual responses in patients with "reactive psychosis" with those of nonpatients. They found smaller amplitudes of the initial components in the patients.

One of the important sources of possible variation between the findings of different investigators arises from the fact that all of the studies so far cited were carried out with mobile pupils. The need for immobilization of the pupils in studies of visual evoked responses is underscored by the results of Rodin and his group. Rodin *et al.* (1964) studied visual responses in a group of schizophrenic patients characterized by a biochemical abnormality reflected in a high lactate–pyruvate *(L/P)* ratio. As indicated in Figure 100, they found that these patients had smaller responses than nonpatient controls and patients with a low *L/P* ratio. Since the patients with low *L/P* ratio tended to have large responses, the responses of Rodin's whole sample of schizophrenics, undivided with respect to *L/P* ratio, would not have been considered different from normal, but the *L/P* ratio provided significant subsamples. However, this interesting finding proved to be artifactual once Rodin *et al.* (1968) repeated their observations in the high *L/P* ratio patients after dilating the pupils with a cycloplegic. The quite different results obtained under these conditions are illustrated in Figure 101. The visual responses of the high *L/P* ratio patients were no longer smaller than normal, and it appeared that the reduced amplitudes observed previously were due to relative miosis of the pupils in these patients. Rodin *et al.* (1968) were also able to show that, with pupils dilated, there were significant correlations between the maximum negativity of the response in the first 250 msec and a number of clinical findings. Low amplitude of the

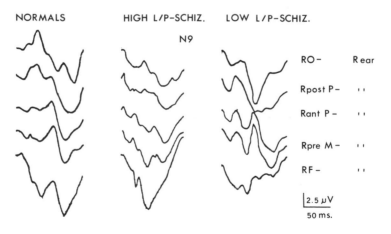

Fig. 100. Comparison of monopolar visual evoked response tracings of 9 nonpatients, and 9 high-*L/P*-ratio and 9 low-*L/P*-ratio chronic schizophrenic patients. The response patterns are shown for the entire right hemisphere. Pupils not immobilized. Note lower-amplitude responses for tracings recorded from posterior portions of the head (top 3 lines) in high-*L/P* patients. (Reprinted with permission from Rodin *et al.*,1968.)

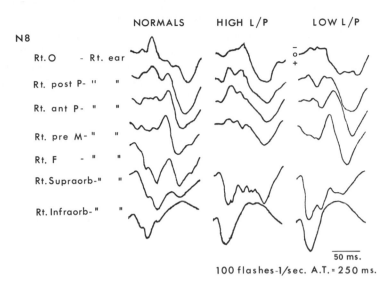

NORMALS   HIGH L/P   LOW L/P

N8

Rt. O   - Rt. ear

Rt. post P- "   "

Rt. ant P- "   "

Rt. pre M- "   "

Rt. F   - "   "

Rt. Supraorb-"   "

Rt. Infraorb- "   "

50 ms.

100 flashes-1/sec. A.T.= 250 ms.

Fig. 101. Tracings obtained in attempt to repeat the study shown in Fig. 100 on another occasion with different normal subjects, but essentially the same group of schizophrenic patients. The pupils of all subjects had been fixed in dilatation for this study. The differences between groups are minimal. (Reprinted with permission from Rodin *et al.*, 1968.)

highest negative peak in the record obtained from the right occipital area was associated with poor nonverbal accessibility, decreased psychomotor activity, decreased amount of speech, absence of sexual interest in student nurses, and poor perception of the body image. Similar, although fewer, relationships were obtained with recordings made from the left occipital area and from the two parietal leads.

The findings of Rodin *et al.* appear to be of considerable importance both from the methodological emphasis that they place on pupillary immobilization, particularly in psychiatric patient groups, and from the fact that they have obtained encouraging correlations with ratings of psychopathology. Since pupillary responses tend to be quite deviant in schizophrenia (Rubin, 1962; Hakerem and Lidsky, 1970), the results of the Rodin group in relation to the state of the pupils are not surprising.

Comment

On the whole, the results obtained by comparing various psychiatric groups with respect to averaged responses to unpaired stimuli, have been either negative or, when positive, difficult to confirm. There seem to be some indications that more consistent results may ensue with greater attention to methodological details, such as immobilization of the pupil and careful stimulus control. However, even the most careful techniques will still leave variables that are very difficult to control, such as the thickness of the tissues intervening between brain and recording electrode. Consequently, the ideal strategy in studies involving unpaired evoked responses may be to use the subject as his own control and to relate changes in the response to psychological changes occurring either spontaneously or with treatment. For purposes of cross-sectional population comparisons, evoked-response indicators that depend upon the relationship between responses rather than upon absolute measurements, do eliminate much of the difficulty introduced by variance resulting from uncontrolled anatomical factors. As will be seen in the succeeding sections, such procedures seem to have yielded more consistent findings in psychiatric studies.

VARIABILITY OF EVOKED RESPONSES IN TIME AND SPACE

The short term variability of evoked responses has been studied by several investigators and related to clinical criteria. The favorite measure of variability has been the product–moment coefficient obtained by correlating corresponding successive data points of two or more average responses. The correlation measure provides a useful index of similarity for comparative purposes, although it does present statistical difficulties (Donchin, 1969).

Callaway's Two-Tone Procedure

Callaway and his colleagues have conducted a series of studies with a procedure in which two tones of 600 and 1000 Hz were randomly intermixed and the averaged auditory response to each was computed (Callaway et al., 1965; Jones et al., 1965; 1966). The subject was told that the test aims at determining how well an individual can ignore tones. Four sets of 40, 600 Hz tones and 40, 1000 Hz tones were presented. The subject viewed his own EEG on a cathode-ray oscilloscope during the procedure. The rationale leading to the original experiments was derived from Shakow's (1963) theory of segmental set. From this theory, it was predicted that schizophrenic patients would be more likely than normals to pay attention to an insignificant difference between stimuli, and that this would lead to differential evoked responses. The results confirmed the prediction; the correlations between the responses to the two tones were significantly lower in schizophrenics than in normals. The correlation measure was also lower in schizophrenics than in patients with neuroses and affective disorders. Figure 102 presents a summary of the results obtained by means of the Callaway procedure. Although there was a trend for the paranoid schizophrenic patients to have higher similarity indices than the nonparanoid patients, this was not statistically significant.

Jones et al. (1966) attempted to relate clinical assessments of a number

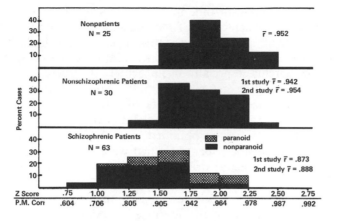

Fig. 102. Distribution of mean two-tone averaged auditory evoked response correlation scores. (Reprinted with permission from Jones *et al.*, 1966.)

of psychopathological variables to the evoked response similarity score. Nurses' ratings of disordered talk and high levels of anxiety tended to be associated with lower two-tone correlations, whereas effective general functioning was associated with high correlations. Patients characterized by incoherent delusional talk tended to have low correlations, whereas high correlations were found in those with coherent organized delusions. Psychiatrists' ratings of disordered thinking were also related to the evoked-response measure. Figure 103 shows the variations in clinical ratings and

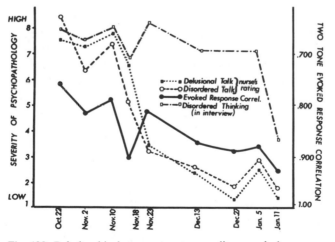

Fig. 103. Relationship between two-tone auditory evoked response measure and certain aspects of clinical change in a typical patient. (Reprinted with permission from Jones *et al.*, 1966.)

evoked-response scores with time in a chronic schizophrenic patient admitted to hospital in an acute state. There is a clear correspondence between clinical improvement, and a tendency toward increasing similarity scores.

Although the two-tone similarity score appears to be significantly lower in schizophrenia, especially in an acutely disordered state, and to be sensitive to clinical change within a given patient, it does not appear to be specific for schizophrenia. Patients with Korsakoff's psychosis and senile dementia also have low two-tone values (Malerstein and Callaway, 1969). In addition, the results obtained with the procedure are subject to the reservation that the patients studied were almost invariably under the influence of psychoactive drugs. However, there is indirect evidence that drug effects on the measure may be minimal, since clinical improvement with no change in drug dosage was accompanied by higher similarity scores. The interpretation of the findings, in relation to the theory of segmental set which generated the procedure, is, however, open to doubt. Jones and Callaway (1970) have recently found that schizophrenic patients were differentiated from normals just as well by the correlation between two averaged responses to identical tones. This means that the significant factor is not the differential response to two tones, but the variability of response. It is still possible that response variability is an outcome of segmental set, but high variability could arise for reasons other than differential attention to insignificant stimuli. For example, loose regulatory control, of the kind proposed for the material on EEG-evoked response amplitude concordance (Chapter 7), could result in increased evoked-response variability.

Visual-Response Variability

The results of the Callaway group, showing increased variability of auditory responses in schizophrenics, appear to be paralleled in the visual modality. Lifshitz (1969) computed the correlations between two average responses to the same visual stimulus in 16 schizophrenic patients and 16 matched controls. The mean correlation between the first and second response was 0.785 for controls and 0.428 for the patients, a statistically significant difference. Lifshitz interpreted the greater variability of evoked responses in schizophrenics to reflect a mechanism responsible for increased uncertainty concerning sensory information. If the brain's electrical representations of sensory perceptions were excessively and randomly variable, this would render certainty about one's perceptions very difficult. Some of the psychopathological phenomena observed in schizophrenia could be accounted for in this way.

Lifshitz (1969) obtained additional evoked-response evidence bearing on defective information processing in schizophrenia. He employed several simple stimuli to synthesize a compound stimulus. He then used statistical methods to attempt identification of the compound stimulus from the evoked response. Such identification was significantly better than chance in nonpatient subjects, but did not differ from chance in schizophrenics. The patient–nonpatient differences appeared to be due to the greater response variability of the schizophrenics. If one regards the experimental identification procedure as a way of interpreting the coding of sensory information, the results agree with Lifshitz's view concerning uncertainty of sensory representation in the brain responses of schizophrenics.

Lifshitz draws attention to the fact that the majority of his patient subjects were receiving medication and that this could have influenced the results. However, five of the schizophrenics had received no medication for six months and he found that the results in this group were not different from those of the other patients.

Ornitz (1969) drew attention to the correspondence between the observations of Callaway and Lifshitz and certain clinical phenomena observed in young schizophrenic children. These children appear to be in a state of perceptual inconstancy with alternating and sometimes almost simultaneous over- and underreactivity to stimuli in all sensory modalities. Ornitz and Ritvo (1968) proposed that this perceptual inconstancy is brought about by inadequate or defective modulation or faulty filtering of sensory input. Evoked-response data, apparently concordant with this interpretation, were obtained by studying auditory evoked responses in children during sleep. In normal children, evoked-response amplitude during the REM phase of sleep is ordinarily reduced relative to that in stage II sleep, particularly during periods of eye-movement bursts. In autistic children, the amplitudes during REM sleep, both in ocular quiescence and with eye movement bursts, were significantly less reduced than in normals. This was interpreted as reflecting decreased phasic inhibition of sensory responses during REM sleep in the autistic children.

Spatial Variability

In addition to being more variable in time, evoked responses in schizophrenic patients, appear to be more variable from one area of brain to another. Rodin et al. (1968) found that the correlations between visual responses recorded from left and right occiput, left and right parietal, and right parietal and occipital areas were lower in schizophrenics than in normals. The correlations between responses from left occipital and left parietal areas did not differ significantly between schizophrenics and nonpatients.

Comment

Thus far, all investigators who have compared evoked-response variability in schizophrenic patients and nonpatients have found greater variability in schizophrenics. The possibility that this finding represents nothing more than a higher "noise" level in the patients, due to a greater admixture of extracerebral activity in the average responses, cannot be completely ruled out. The possible influence of psychoactive drugs also clouds some of the findings. However, with these reservations in mind, it would appear that the measurement of variability may offer a meaningful approach to evoked-response analysis in psychopathological states.

SOMATOSENSORY EVOKED-RESPONSE RECOVERY FUNCTIONS

Early Studies

Our first study of somatosensory recovery in psychiatric patients compared a heterogeneous sample of 92 patients with 13 nonpatients (Shagass and Schwartz, 1961*b*, 1961*c*, 1962*a*). Figure 104 shows the mean recovery curves for the initial negative–positive component. The main difference between the groups was in the portion of the curve involving

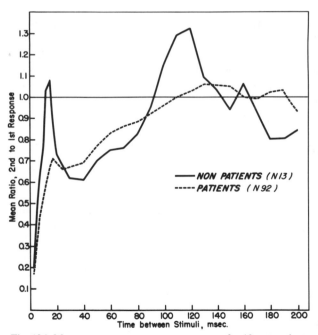

Fig. 104. Mean somatosensory recovery curves for 13 nonpatients and 92 patients of various kinds. All measurements in this and subsequent figures for somatosensory recovery refer to the initial negative–positive component unless otherwise designated. Note early peak of recovery occurring before 20 msec interval in nonpatients and not in patients. (Reprinted with permission from Shagass and Schwartz, 1961*b*.)

interstimulus intervals from 2.5 to 20 msec. Whereas the peak mean ratio for the nonpatients exceeded unity, that for the patients was well below this level. A second statistically significant difference was found at about 100 msec, but this was less consistent than that for the early portion of the curve.

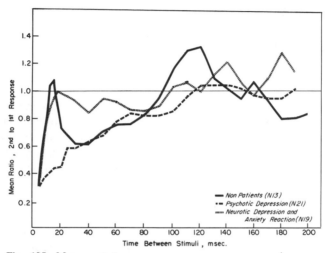

Fig. 105. Mean somatosensory recovery curves comparing non-patients, psychotic depressives and dysthymic neuroses. Psychotic depressives differ from the other two groups. (Reprinted with permission from Shagass and Schwartz, 1961c.)

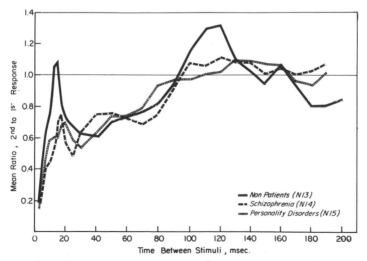

Fig. 106. Mean somatosensory recovery-function curves comparing nonpatients with schizophrenics and personality disorders. (Reprinted with permission from Shagass and Schwartz, 1961c.)

Breaking down the patient sample into several major diagnostic categories, it appeared that the only clinical group that did not differ significantly from nonpatients with respect to recovery function was the neurotic depression and anxiety reaction (dysthymic) group. Figure 105 shows the mean curves comparing nonpatients with dysthymics and psychotic depressions. The values during the first 20 msec were clearly lower in the psychotic depression group than in the other two. As seen in Figure 106, the diminished early recovery was also present in patients with diagnoses of schizophrenia and personality disorders.

Personality Disorders

The deviant somatosensory recovery function findings in personality disorders were of special interest, since there are two opposed views concerning these disorders. Mayer-Gross *et al.* (1954) have grouped them with the psychoneuroses in their textbook, whereas Cleckley (1959) considers them to be variants of psychosis. From the point of view of the somatosensory recovery function findings, the personality disorders appear to resemble psychosis more than psychoneurosis. This was supported by additional data in which 36 patients with various personality disorders were compared with 25 nonpatients, 34 patients with psychoneuroses, and 21 schizophrenics (Shagass and Schwartz, 1962*b*). Figure 107 shows the distribution of the peak recovery ratios obtained during the first 20 msec in the various groups. The distribution of the personality disorder group is clearly like that of the schizophrenias and different from those of the

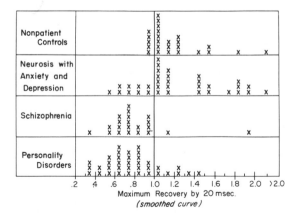

Fig. 107. Distributions of peak somatosensory recovery ratios, obtained with 8 interstimulus intervals of 20 msec or less, in four groups of subjects. Peak ratios are lower in schizophrenics and patients with personality disorders. (Reprinted with permission from Shagass and Schwartz, 1962*b*.)

other groups. Comparisons of the various diagnostic categories within the personality disorder group showed only that the results in patients diagnosed as "sociopathic personality disturbance, antisocial reaction" were more homogeneous than those of the others. The greater physiological homogeneity would be in accord with the greater clinical homogeneity of this group.

Confirmatory Studies

Findings in our initial studies were expanded and confirmed in a subsequent report which added 27 nonpatients and 75 patients of various kinds to the initial sample of 13 and 92 subjects, respectively (Shagass and Schwartz, 1963c). Once again, the peak recovery ratios during the first 20 msec were significantly higher in the nonpatients and, within the patient population, the findings in the dysthymic group approached those of the nonpatients. Figure 108 shows the distribution of peak ratios by 20 msec for 40 nonpatients, 94 patients with personality disorders, schizophrenias, or psychotic depressions, and 73 other patients. The distribution of the personality disorder, schizophrenia, and psychotic depression group is clearly different from that of the other two, their average ratio being much lower. About half of the other patient group was composed of dysthymics. Among the remaining patients, the following results are noteworthy: six patients with brain syndromes gave the highest median ratio of the entire sample; the values for eight undiagnosed patients and five with paranoid states approached those of the nonpatients; six patients with conversion reactions and three with manic states had a preponderance of low ratios.

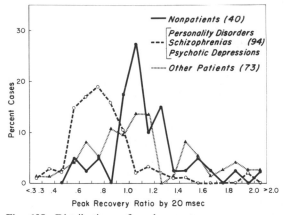

Fig. 108. Distributions of peak somatosensory recovery ratios (8 intervals from 2.5 to 20 msec) in three groups of subjects. Results for patients with personality disorders, schizophrenias and psychotic depressions were grouped together because they were similar. (Reprinted with permission from Shagass and Schwartz, 1963c.)

In our study of somatosensory intensity–response relations (Shagass and Schwartz, 1963a), we also measured recovery functions in 44 patients. The correlation between the intensity gradient slope and the peak recovery ratio by 20 msec was -0.37 ($p < 0.02$). To determine whether the diagnostic differentiations provided by the slope and recovery measures were the same or different, the results in 24 patients were analyzed. Twelve patients with diagnoses of schizophrenia, psychotic depression, or personality disorders were compared with twelve other diagnoses. The slope and recovery measures both differentiated between the groups at the 0.01 level, while R_1 amplitude did not discriminate. Since the correlation between the slope and recovery ratio became nonsignificant when amplitude of R_1 was partialled out, indicating that amplitude was the common factor, it appeared that the intensity slope and recovery function were independent correlates of psychiatric diagnosis.

Although age was not significantly correlated with the recovery ratio, we did observe a sex difference. Peak recovery by 20 msec was significantly lower in females than in males. The differences between patients and non-patients, however, remained significant when they were compared within each sex group. Furthermore, the nonpatients did not show the sex difference; it appeared to be present only in patients. It was thought to be attributable to unequal sex distribution by diagnosis. Despite the statistical assurance that the patient–nonpatient differences were not due to factors unrelated to psychopathology, the validity of the findings was probably best supported by evidence that clinical improvement was accompanied by normalization of recovery values.

Serial Studies in Psychotic Depressions

Psychotic depression is perhaps the psychiatric condition best suited for longitudinal study in relation to clinical change, since there are effective treatments and the therapeutic modifications occur within a relatively short time. We were able to obtain serial recovery function measurements in 14 psychotic depressive patients before and during treatment (Shagass and Schwartz, 1962c). Most of the patients received ECT. Figure 109 shows the mean recovery curves obtained before treatment, toward the end of treatment when some degree of symptomatic improvement had occurred, and just prior to discharge from hospital. The mean peak ratios during the first 20 msec show a progressive increase from an initial low to a maximum in the last test. The results of three individual patients are shown in Figure 110. They illustrate the fact that the group mean recovery curves are attenuated because the peak value occurs at different interstimulus intervals in different subjects. The patient whose scores are shown at the right of Figure 110 had recovery values approximating the norm prior to treatment, but nevertheless showed an increased recovery ratio after successful therapy. Additional evidence that peak recovery varies in accordance with

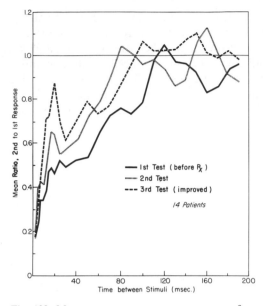

Fig. 109. Mean somatosensory recovery curves for 14 psychotic depressive patients before, during, and after treatment. Note progressive increase of early peak recovery ratio. (Reprinted with permission from Shagass and Schwartz, 1962c.)

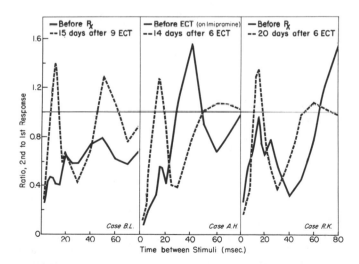

Fig. 110. Recovery curves for first 80 msec of 3 patients before and after treatment with ECT. (Reprinted with permission from Shagass and Schwartz, 1962c.)

clinical state is presented Figure 111 which shows the peak ratios for a patient who was studied during several remissions and relapses. The ratios correspond to the patient's clinical condition, being high when she appeared in remission and low when she relapsed. Although most of our patients were treated with ECT, the normalization of recovery values did not appear to depend on this particular therapy. Figure 112 shows serial recovery curves obtained in a patient treated with imipramine alone; the results are like those obtained with ECT.

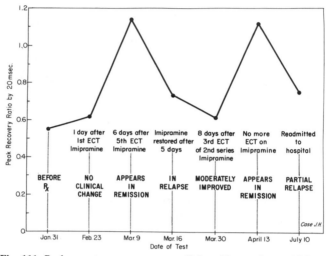

Fig. 111. Peak somatosensory recovery (2.5 to 20 msec intervals) in a psychotic depressive patient undergoing remissions and relapses. (Reprinted with permission from Shagass and Schwartz, 1962c.)

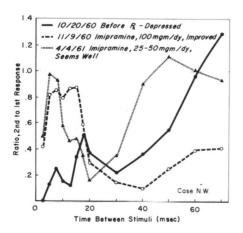

Fig. 112. Recovery curves before and during treatment with imipramine in a manic-depressive, depressed patient. Early recovery ratio increases in association with clinical response. (Reprinted with permission from Shagass *et al.*, 1962.)

ADDITIONAL STUDIES OF SOMATOSENSORY RECOVERY FUNCTIONS

Our most recent large-scale study of somatosensory recovery functions involved 178 psychiatric patients and 89 nonpatient controls (Shagass, 1968a). Several methodological features that distinguished this study from previous ones have been outlined in the section on unpaired somatosensory responses earlier in this chapter. One additional technical aspect was the use of R_2 values adjusted by covariance, in parallel, in order to avoid the statistical problems presented by the ratio (Chapter 2).

Sixteen of the 178 patients had chronic brain syndromes and were 65 years of age or older; these were dealt with separately. The recovery data of the remaining 162 patients were compared with those of 54 nonpatients matched for age and sex. There were eight amplitude and latency measurements in addition to a composite amplitude measurement between peaks 1 and 4. Figure 113 shows the mean recovery curves for the six amplitude peaks that yielded statistically significant differences. With the exception of peak 1, the controls showed greater positivity, particularly during the first 20 msec, for all peaks. Since peak 1 is clearly negative, the greater negativity in nonpatients also indicated greater recovery. Although peaks 5 and 7 are theoretically negative, they are often seen above the base line and the greater positivity in the controls probably suggests greater recovery. However, this could really mean greater recovery of the adjacent positive peaks (4, 6, and 8) in the controls, rather than representing a measurement of the negative values. Figure 114 shows the latency-recovery measurements that discriminated significantly between nonpatients and

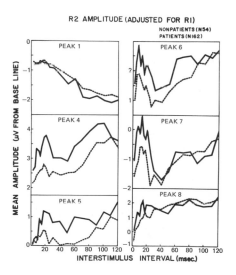

Fig. 113. Mean somatosensory amplitude-recovery curves of 162 patients (broken line) and 54 nonpatients (solid line) matched for age and sex. There were significant differences in at least one segment of all 6 curves. (Reprinted with permission from Shagass, 1968a.)

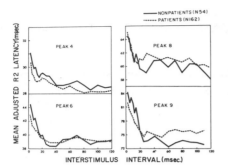

Fig. 114. Mean somatosensory latency recovery curves of 162 patients and 54 nonpatients matched for age and sex. There were significant differences in at least one sigment of all 4 curves. (Reprinted with permission from Shagass, 1968a.)

patients. Latency recovered more rapidly in the patients for peaks 4, 6, and 9. This suggests that latency and amplitude represent different measures of recovery.

Effect of Medication

One of the important issues in evoked-response studies of psychiatric patients concerns the length of time that the patient should be free from drugs before the results can be accepted as not reflecting a pharmacological influence. To obtain statistical evidence that would bear on this point, 39 patients who had no history of receiving medication were matched for age and sex with 39 patients who had. The analysis of variance showed three R_1 differences, each at $p < 0.05$, and two R_2 interactions between age and medication, also at $p < 0.05$. Since this number of significant differences could have occurred by chance, it was concluded that no differences associated with medication were demonstrated. Furthermore, those differences that were found had no apparent relationship to significant patient–nonpatient differences.

Comparisons of Major Diagnostic Groups

Additional analyses were conducted to determine whether evoked-response differences were specifically related to major classes of psychiatric illness, i.e., psychosis, psychoneurosis, and personality disorder. In one analysis, all patients with a diagnosis of schizophrenia, psychotic depression, paranoid reaction, or manic state were compared with nonpsychotic patients (psychoneuroses, personality disorders, and adjustment reactions) and nonpatients who could be matched for age and sex; there were 39 subjects in each group. Virtually all significant differences were found to occur between patients and nonpatients. The only measure that discriminated between psychotics and nonpsychotics was the adjusted R_2 for peak 1–4 in the first 20 msec of the recovery curve; the mean value for nonpsychotics was intermediate between those for nonpatients and psychotics

(Figure 115). This result appeared to provide some confirmation of earlier data indicating that recovery findings were more like normal in dysthymics but, as indicated below, further analysis did not substantiate this conclusion.

In another set of analyses, comparisons were made between nonpatients, patients with psychoneuroses, and patients with personality disorders matched for age and sex; there were 36 subjects in each group. Again, the differences between patients and nonpatients were responsible for all but one of the significant F ratios; the adjusted R_2 for latency 9 was significantly greater for psychoneurotics than for personality disorders for one segment of the recovery curve. The results in these two analyses were, on the whole, disappointing, since they yielded almost no differentiations between patient groups.

Individual Diagnoses

Further analysis in relation to diagnosis had to be confined to comparisons of individual patient groups with nonpatients matched to them for age and sex, because age and sex distributions of the patient groups differed from one another. Table IV lists the measures that yielded significant differences in the patient–nonpatient comparisons. It will be seen that

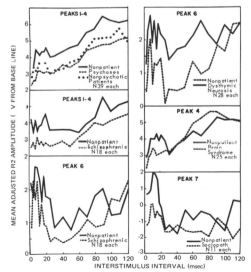

Fig. 115. Mean somatosensory amplitude-recovery curves for selected peaks illustrating differences between various patient groups and nonpatients matched for age and sex. (Reprinted with permission from Shagass, 1968a.)

Table IV. Somatosensory Evoked-Response Measures Yielding Significant Differences ($p < 0.05$) in Comparisons of Patient Groups with Nonpatients Matched for Age and Sex

Patient group	Amplitude		Latency	
	R_1	Adjusted R_2	R_1	Adjusted R_2
Psychoneurosis, dysthymic ($N = 28$)		$1^c, 4^c, 5^b, 6^b, 7, 1$–4	7	$5^c, 8^c$
Psychoneurosis, conversion ($N = 14$)				
Emotionally unstable ($N = 13$)		10^c		10^c
Passive–aggressive ($N = 13$)	1–4	$5, 7^c$	10	$1^c, 6^c, 7$
Sociopathic, antisocial ($N = 11$)		$6, 7^b$		1^c
Brain syndrome ($N = 25$)		$1^c, 4^c, 10$		6
Schizophrenia ($N = 18$)	10	$4, 5^c, 6, 7, 1$–4	5	$1^c, 4^a, 5$
Psychotic depression ($N = 21$)		$1^c, 4^b, 7^c, 1$–4^b		$6^c, 7^c$

[a] $p < 0.01$.
[b] $p < 0.001$.
[c] Significant only in segment of curve.

the largest number of measurements providing significant differences was found for the comparisons involving dysthymic psychoneuroses, schizophrenias, and psychotic depressions. However, since these patient groups were also among the largest, it appeared that sample size was a major determinant of statistical significance. Examination of the mean recovery curves confirmed this, since the differences between most patient groups and the matched nonpatient group were similar in degree and direction. Adjusted R_2 amplitude differences were nearly always in the direction of greater recovery in nonpatients; adjusted R_2 latency differences were nearly always in the direction of slower recovery in nonpatients. Figure 115 illustrates some of the mean amplitude-recovery differences for several diagnostic groups. The results for adjusted R_2 amplitude 7 in sociopathic personality disturbance were of special interest because there were few other differences between this group and nonpatients, while this one measure provided a striking discrimination. Using the mean value for the first 20 msec as an index, only one of 11 subjects, in each group, had a value overlapping those of the other group.

Figure 116 shows the mean amplitude recovery curves for peaks that distinguished psychotic depressions from their matched controls (Shagass and Schwartz, 1966). Another way to illustrate the main finding, which is that R_2 of peak 4 was greater in nonpatients for any given R_1 value, is to plot the regression of R_2 on R_1 across subjects. This is done, using the mean value for eight interstimulus intervals from 2.5 to 20 msec, in Figure 117. It will be seen that, for any value of R_1, the probability that R_2 would be lower in the psychotic depressives than in the controls was quite high.

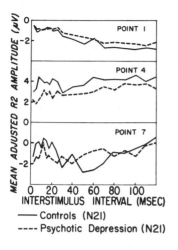

—— Controls (N21)

- - - - Psychotic Depression (N2I)

Fig. 116. Mean somatosensory amplitude-recovery curves comparing psychotic depressive patients and nonpatients matched for age and sex. There were significant differences between the groups for at least one segment of the curves. (Reprinted with permission from Shagass and Schwartz, 1966.)

SOMATO SENSORY RECOVERY-PEAK 4

8 MEASURES FROM 2.5 TO 20 msec

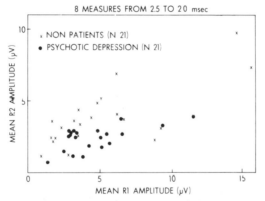

Fig. 117. Scattergrams obtained by relating mean R_2 amplitude (unadjusted) for 8 interstimulus intervals from 2.5 to 20 msec to mean R_1 amplitude for subjects of Fig. 116. Each X or dot is one subject. Note that for any given value of R_1, R_2 tends to be greater in nonpatients.

Figure 118 shows a similar plot for the schizophrenic patients and their matched controls.

Multiple Regression Analysis

Since the various evoked-response measurements were correlated with one another to varying degree, a multiple regression analysis was performed to determine the optimal combination of variables that would predict the patient–nonpatient criterion. In this analysis, mean R_1 amplitude and

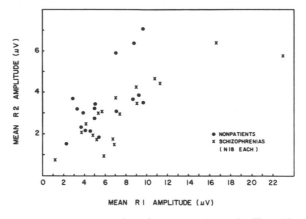

Fig. 118. Scatterplot of the type shown in Fig. 117, comparing nonpatients with schizophrenic patients matched for age and sex.

latency values for eight peaks across all interstimulus intervals, and mean adjusted R_2 values for each of three segments of the recovery curve, provided the evoked-response variables. The three segments covered the interstimulus intervals from 2.5 to 20, 25 to 60, and 70 to 120 msec. There were thus 64 evoked-response variables. Together with age and sex, they were tested as predictor variables against the patient–nonpatient criterion. With all 66 variables, the initial multiple correlation was 0.685. After successive rejection of noncontributory variables, the final multiple correlation between 16 residual evoked-response variables and the criterion was 0.590 ($p < 0.001$). Age and sex were eliminated as noncontributory. Clearly, many of the evoked-response variables were redundant. Furthermore, it is of interest that only six of the 16 final variables were, by themselves, significantly correlated with the criterion. The retention of the remaining ten in the final regression equation was based upon the nature of their intercorrelations with other variables. The elimination of several variables that gave significant criterion correlations by themselves indicates that they replicated those retained. The single variable that correlated most highly with criterion was adjusted R_2 amplitude 4 for the 2.5 to 20 msec intervals. The point biserial coefficient for this variable was 0.293 which, although highly significant, indicates a reduction of predictive error from chance of no more than 8.6%.

Comment

The main way in which the results of the latest somatosensory recovery study disagreed with earlier findings was in showing that dysthymics differed from normals in the same manner as patients with "functional" psychoses and personality disorders. This leaves the results with virtually no diagnostic specificity; "patienthood" appears to be associated with diminished amplitude recovery and earlier latency recovery, but type of "patienthood" seems to make little difference. Although the demonstration that peak 1–4 recovery is greater during the first 20 msec in nonpsychotics than in psychotics suggests that there may be some diagnostically related differences within the patient group, these did not appear to be very striking.

The measurements of different evoked-response peaks tended to provide the same kinds of diagnostic differentiations (Figures 113 and 114). Thus, there also appears to be a relative absence of specificity in relation to diagnosis across the various evoked-response measures. In view of the difficulty in accurately measuring amplitudes and latencies of peaks occurring after the first component (Chapter 2), it should perhaps not be surprising that there was little diagnostic differentiation between separate measures of each kind. However, the empirical demonstration that this was so has served as a guide to us in subsequent studies, in the sense that we have accepted as a reality that there is little to be gained from attempts to identify somatosensory peaks after the initial negative–positive component with high accuracy. The low degree of specificity, both in diagnostic relations and in discriminations provided by different evoked-response measurements, suggested that it might be more profitable to explore variations of the recovery function procedure as a means of gaining new diagnostic correlates, than to devote special efforts to peak identification.

MODIFIED SOMATOSENSORY RECOVERY FUNCTION PROCEDURE

The somatosensory recovery function studies described above all employed conditioning and test stimuli of equal intensity. The intensity selected was generally on the asymptote of the intensity–response curve, since we wished to produce a large amplitude response and to reduce variability. Early in our studies, however, we had noted that degree of recovery was very markedly influenced by stimulus intensity. Usually recovery was greater when stimulus strength was lower (Shagass and Schwartz, 1962d). When we measured recovery functions with different stimulus intensities in the same subject, we found that the within-subject correlation between evoked-response amplitude and peak recovery was about the same as the correlation between subjects with a single intensity. The lower the amplitude of R_1, whether due to weak stimuli or to the amplitude limit of the subject in response to high-intensity stimuli, the greater the recovery. In some subjects, there was marked "facilitation" when recovery was measured with low-intensity stimuli and marked suppression of recovery when high-intensity stimuli were employed (Figure 31). This suggested the possibility that, by varying stimulus intensity, the recovery function procedure could be used to measure both "facilitatory" and "inhibitory" cortical reaction tendencies. Our recent experimental work has been devoted mainly to exploration of the psychiatric correlates of a procedure in which conditioning stimulus intensity is varied.

Another modification introduced into the procedure was to apply trains of nine conditioning stimuli in addition to the usual single one. It was reasoned that, if the reduced recovery found in association with psychopathology actually reflects impairment of some restorative mechanism that comes into play after neuronal excitation, such impairment should be more manifest after several closely spaced conditioning stimuli than following a single one. Some of the preliminary results obtained with this modified recovery function procedure will be presented here.

Method

Within each averaging sequence, five different stimulus configurations are presented in pseudorandom order by means of a programming device. The five configurations are as follows: (1) single stimuli of test intensity; (2) single stimuli of conditioning intensity; (3) pairs, composed of single conditioning and single test stimuli; (4) trains, composed of nine conditioning stimuli and a single test stimulus; (5) trains of nine conditioning stimuli. Four responses are obtained in each averaging sequence. Utilizing subtraction, estimates of the response to the test stimulus following a single conditioning stimulus (R_2) and the response to the test stimulus following the train of nine conditioning stimuli (R_{10}) are obtained in separate channels of the computer. R_2 is obtained as before by subtracting R_1 from $R_1 + R_2$. R_{10} is obtained by subtracting the responses to the train of nine from those to the train of ten. The response to the single unconditioned stimulus of test intensity ($R_1 T$) is also obtained to provide a base-line estimate against which to judge the effect of the conditioning stimuli upon the test responses. The fourth averaged response is that to the conditioning stimulus alone ($R_1 C$). A single interval between stimuli is employed, namely, 10 msec. This interval was selected after a pilot study in which eight subjects were tested repeatedly in order to measure full recovery functions with various conditioning stimulus intensities. Figure 119 shows the results in one of these subjects. It will be seen that in this subject, as in others, a 10 msec interval was quite representative of the effects of varying conditioning-stimulus intensity on degree of recovery.

There are nine averaging sequences in the experiment. Five are performed with test stimulus intensity 10 ma above threshold; the conditioning stimulus intensities are: 10, 5, and 2 ma above threshold, threshold, and 0.5 below threshold. In four sequences, the test stimulus intensity is 5 ma above threshold; conditioning stimulus intensities are: 10, 5, and 2 ma above threshold and threshold.

Figure 120 shows sample average responses from one subject, obtained with test stimuli 10 ma above threshold; in one sequence, the conditioning stimulus was also 10 ma above threshold; in the other, the conditioning

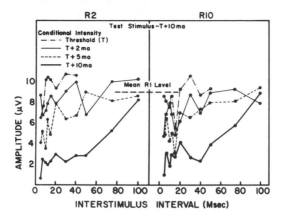

Fig. 119. Somatosensory recovery-function curves for one subject studied in 6 experimental sessions. Conditioning-stimulus intensity was varied, while test stimulus was always 10 ma above sensory threshold. Responses with single conditioning stimuli and trains were obtained in the same averaging sequence. Test response amplitude was generally greater with lower-intensity conditioning stimuli. Differences between the curves appear to be well represented at an inter-stimulus interval of about 10 msec.

stimulus was of threshold intensity. Whereas R_2 was of much lower amplitude than the R_1T stimulus with the high-intensity conditioning stimulus, it was larger than R_1T when it was preceded by a threshold conditioning stimulus. R_{10} was of lower amplitude than R_1T with both conditioning intensities, although it was larger following the threshold train than after the high-intensity train. Our experience so far indicates that "facilitation" of the initial R_2 negative–positive component, as illustrated in the lower set of tracings in Figure 120, does not always occur when the test stimulus is 10 ma above threshold. Only about half of nonpatient subjects have shown greater R_2 or R_{10} than R_1T with threshold conditioning-stimulus intensities. With a test stimulus intensity 5 ma above threshold, R_2 and R_{10} are more often greater than R_1T. There may be a limited capacity for amplitude increase beyond the level reflected in the R_1T when stimulus strength is 10 ma above threshold.

Some Preliminary Results

Although we have accumulated considerable data employing this procedure, only small portions have so far been analyzed. These deal only with the time epoch from 15 to 31 msec after the stimulus and with the computer measurement of average deviation around the epoch mean. The

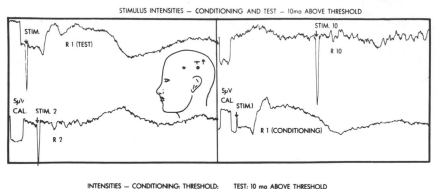

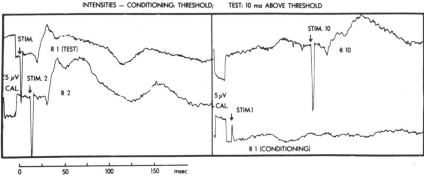

Fig. 120. Sample tracings from one subject obtained in modified somatosensory recovery function procedure. Upper tracings, both conditioning and test stimulus intensities were 10 ma above threshold. R_1 (test) and R_1 (conditioning) each represent averages to 50 single stimuli of test and conditioning intensity; they provide an indication of reliability of the average. R_2 is the response to the second stimulus of a pair; R_{10} is the response to the tenth stimulus of the train. Interstimulus interval, 10 msec. Note small amplitude of both R_2 and R_{10} with high-intensity conditioning stimulus. In lower tracings, conditioning-stimulus intensity was set at threshold; this is reflected in very low amplitude of R_1 (conditioning). R_2 is of higher amplitude than R_1 (test), but R_{10} still shows relative diminution of initial component amplitude.

average deviation is highly correlated with computer and hand peak-to-peak measurements of the initial negative–positive component amplitude. Our analyzed data are also confined to the five averaging sequences in which the test stimulus is 10 ma above threshold.

Two evoked response indices related to psychopathology were obtained from an analysis comparing 10 nonpatients with 11 schizophrenics and 11 nonpsychotic patients (Shagass *et al.,* 1969). Although about equally matched for sex, the patient groups were older than the nonpatients. One indicator was the mean amplitude of R_{10} across the five stimulus conditions employed. Figure 121 shows the scatter plots relating mean R_2 and R_{10} amplitude to that of $R_1 T$. Although the groups seemed completely intermingled with respect to R_2, there was a clear trend for the amplitude of R_{10} to be lower in schizophrenics for any given amplitude of $R_1 T$. The data were analyzed by computing the regression equation for the entire group of subjects and determining the deviation of the actual R_{10} from that predicted for each subject from the regression equation. An analysis of variance showed these deviation values to be significantly different between groups. Both the schizophrenics and the nonpsychotic patients differed from the nonpatients.

The fact that mean R_{10} discriminated between groups while R_2 did not was encouraging. The absence of R_2 discrimination could possibly be attributed to the fact that schizophrenics were older than controls and would therefore be expected to have greater recovery (Chapter 4). Also, only one interstimulus interval was used. The significantly lower R_{10} in the schizophrenics suggested that application of a train did indeed bring out greater recovery differences than when only one conditioning stimulus was used.

The second indicator of interest was obtained by ranking the amplitude of the response in relation to conditioning-stimulus intensity. Rank 1 indicated the highest amplitude while rank 5 was assigned to the lowest. Figure 122 shows the mean ranks for each of the four types of average response. The curves for the three groups were quite similar with respect to $R T$ and the response to single conditioning stimuli ($R_1 C$). However, for R_2, they differed significantly; whereas the nonpatients had a regular

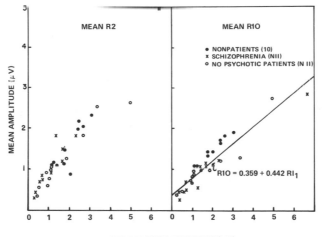

MEAN RI (TEST) AMPLITUDE (μV)

Fig. 121. Scattergrams in which the mean amplitude, in five averaging sequences, of R_2 and R_{10} are plotted against the mean amplitude of R_1 (test). Test-stimulus intensity, 10 ma above threshold. Conditioning stimulus intensities: 10, 5, and 2ma above threshold, threshold, and 0.5 below threshold. Note linear relationships. The three groups of subjects were not differentiated by mean R_2, but for a given value of mean R_1 (test), mean R_{10} was higher in nonpatients than in patients. (Reprinted with permission from Shagass *et al.*, 1969.)

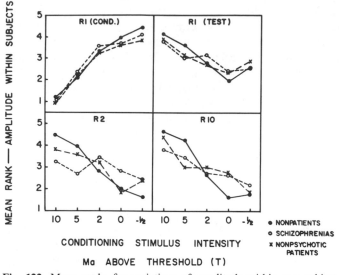

Fig. 122. Mean ranks for variations of amplitude within one subject of R_1 (test), R_1 (conditioning), R_2, and R_{10} as a function of differing conditioning-stimulus intensity. Rank 1 indicates highest amplitude: rank 5, lowest. Note that schizophrenic patients have almost no gradient for R_2 whereas gradient is steep in nonpatients. (Reprinted with permission from Shagass *et al.*, 1969.)

gradient, with highest R_2 amplitude following the threshold conditioning stimuli and lowest R_2 amplitude following the most intense conditioning stimuli, the schizophrenic group showed virtually no gradient. The schizophrenics also had a less steep gradient for R_{10}, but the difference was not statistically significant. This finding suggested a disturbance in "excitation-inhibition" balance in the schizophrenics. The absence of group differences in R_1T and R_1C indicates that the R_2 gradient differences were not due simply to greater variability in the schizophrenics.

The data of a more recent study raise the possibility that the results may indicate something other than recovery (Shagass *et al.*, 1971). Findings were analyzed in patients requiring psychiatric hospitalization because of reactions to drugs. Patients with a history of drug abuse were divided into two groups: those with a history of psychotic reactions and those without such a history. Their results were compared with those obtained in nonpatients and schizophrenic patients of the same age. The relevant analysis was performed only on male subjects because there were marked differences between males and females within the psychotic patient groups. For each conditioning stimulus, the values of R_2 and R_{10} were adjusted for the accompanying R_1T by means of the within-group regression equation. The adjusted values were then subjected to a "mixed" design analysis of variance (repeated measures and diagnosis).

The analysis showed lower mean R_1C, R_2, and R_{10} values in the psychotic drug user and schizophrenic groups than in the nonpatients and nonpsychotic drug users (Figure 123). The most striking difference between the groups seen in Figure 123 is for the gradients of R_2 and R_{10} with respect to conditioning-stimulus intensity. The R_2 and R_{10} curves of the psychotic drug users and the R_2 curve of the schizophrenics appear relatively flat when compared with those of the other two groups. This difference in curve shape is reflected in significant interaction terms (groups × conditioning-stimulus intensity) found in the analysis of variance. However, if one brings the curves together, as in Figure 124, it appears that the significantly lower overall adjusted R_2 and R_{10} values of the psychotic groups cannot be interpreted as reduced recovery with all conditioning-stimulus intensities. It is true that recovery was less than that of other groups in psychotic drug users when conditioning stimuli were about threshold. However, the mean R_2 of the psychotic drug users was actually greater than their mean R_1T for the highest conditioning stimulus intensity, whereas the R_2 in controls and nonpsychotic drug users was smaller than R_1T.

The data can be interpreted as indicating a generally reduced level and range of responsiveness in the psychotic drug users and schizophrenic patients. Whereas the responses of the nonpatients and nonpsychotic drug users displayed a considerable dynamic range as a function of stimulus conditions, those of the psychotic groups did not. Such a reduction of range of responsiveness seems to reflect a tonic state of "inhibition" of cortical responses to external stimuli.

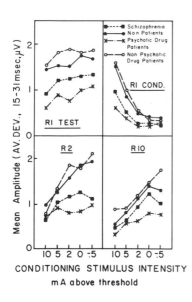

Fig. 123. Mean curves for four groups of subjects in modified somatosensory recovery-function procedure. Test stimulus intensity, 10 ma above threshold. All subjects were male. There were 9 schizophrenics, 15 nonpatients, 8 psychotic drug patients, and 6 nonpsychotic drug patients. Note relatively flat R_2 curves for psychotic drug patients and schizophrenics.

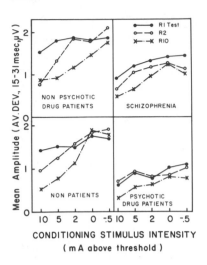

Fig. 124. Mean curves from Fig. 123 rearranged to show differences in intercurve relationships between various subject groups.

Comment

Two main conclusions can be drawn from these early findings with the modified recovery function procedure. One is that the results provide further evidence that psychopathology is associated with altered relationships between evoked-response events. From this point of view, the results agree with those obtained by measuring evoked-response variability and conventional recovery functions. The other conclusion is that studies involving manipulation of stimulus parameters offer a potentially high reward in the application of evoked-response methods to psychiatric problems. Although measurement of responses to single stimuli is far easier, the "payoff," in terms of correlates that may lead to pointed experimental questions, appears to be greater with the more complex parametric procedures.

AUDITORY EVOKED-RESPONSE RECOVERY FUNCTIONS

The only report of auditory recovery function studies in psychiatric patients is that of Satterfield (1969*a*, 1969*b*). He directed his attention to the negative–positive wave (N_1–P_2) occurring after 100 msec. His technique for measuring recovery was to use trains of stimuli at four different rates ranging from 1/sec to 1 every 8 sec. The measure of recovery was the ratio between the amplitude of the response at the fastest rate and that at the slowest. Twenty-two depressed patients were compared with 22 matched nonpatients.

Prior to treatment, mean amplitudes and percent amplitude recovery were greater in the depressed patients than in the controls, but these differences were not significant. The distribution of recovery percentages was broader in the patients than in the controls with patients falling outside of the normal distribution at both low and high ends. Satterfield then compared the two groups with extreme high and low recovery. The finding of greatest interest was that the patients with low recovery gave a history of depressive illness in a first-degree relative, whereas none of the high-recovery patients had a positive family history.

The effects of ECT were not statistically significant; there was a tendency for regression toward the mean, with reduced recovery in the patients who had had high recovery and increased recovery in those who had had low recovery before treatment.

Satterfield's findings with auditory responses appear to be at some variance with the results obtained employing somatosensory and visual modalities. However, he was looking at an evoked-response event that is later and not comparable to those measured by investigators of the other types of response. Consequently, the results are not actually contradictory, but rather suggestive of the possibility that different relationships with psychopathology may await exploration of the recovery functions of later components of visual and somatosensory responses. Satterfield's data suggesting that recovery differences may be related to the presence or absence of a family history of affective disorder point to the possible value of family history of mental illness as a clinical criterion variable.

VISUAL EVOKED-RESPONSE RECOVERY FUNCTIONS

Our study of visual recovery functions involved 78 psychiatric patients and 19 nonpatient controls. Three peak-to-peak amplitude measurements and four determinations of latency were made (Figure 43). The R_2 values were adjusted by covariance in parallel for each of 23 interstimulus intervals ranging from 5 to 200 msec. Statistical analysis was done separately for the 13 intervals from 5 to 100 msec and for the ten from 110 to 200 msec. The subject population was divided in three different ways: (1) nonpatients–patients; (2) nonpatients–psychotic patients–nonpsychotic patients; (3) nonpatients–psychoneuroses–personality disorders–schizophrenia–other psychoses.

The analyses for intervals from 5 to 100 msec yielded significant differences only for latencies 1 and 3. All of these showed slower latency recovery in patients than in nonpatients. The patient groups as a whole had delayed recovery of latency 1, and the nonpsychotic and psychotic groups were both significantly different from the nonpatients. While the psychotic groups as a whole also showed prolonged recovery of latency 3, this difference appeared to be contributed more by the nonschizophrenic than by the schizophrenic patients. No significant differences were obtained in the analysis for interstimulus intervals from 110 to 200 msec.

In contrast to our own negative results with amplitude recovery of the visual response, Speck *et al.* (1966) obtained clear evidence that recovery

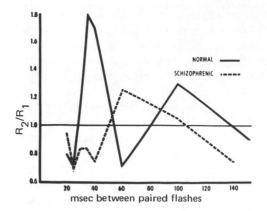

Fig. 125. Visual evoked response recovery function curves for negative peak 3 (latency 60 to 90 msec). Note absence of initial facilitation in schizophrenic subject. (Reprinted with permission from Speck *et al.*, 1966.)

was reduced from normal in psychiatric patients. Their main data were obtained from measurements of negative peak 3, with latency ranging from 70 to 90 msec. Figure 125 shows typical recovery curves for a nonpatient and for a schizophrenic patient. Although shifted somewhat to the right, the curves resemble those found by us for the somatosensory recovery function with an early peak of "facilitation" present in the nonpatient and not in the schizophrenic. Speck *et al.* found that 80% of their nonpatients showed a supernormal peak around the 30 to 35 msec interval. A recovery ratio obtained by comparing responses to paired flashes at an interstimulus interval of 35 msec with single-flash responses showed greater ratios in the nonpatients ($p < 0.001$). More than half of the patients studied by Speck *et al.* were classed as schizophrenics, but there were 16 patients with personality disorders and a few manic-depressives and psychoneurotics. The patient groups did not differ significantly from one another, an absence of diagnostic specificity that parallels our results with somatosensory recovery.

Heninger and Speck (1966) related a number of clinical ratings made before and during the course of treatment to changes in visual evoked-response variables. Of particular interest here is the fact that their recovery-ratio measure was significantly correlated with clinical improvement. With improvement the ratio shifted toward nonpatient values. The amount of change in the ratio was correlated with degree of improvement in: motor activity, degree of involvement with the examiner, amount of affect, anxiety, and hallucinations. The symptom that showed the greatest improvement with treatment (hallucinations) had the highest correlation with an increase in the recovery ratio.

Deviant visual recovery functions in schizophrenia have also been found by Floris *et al.* (1967, 1968) in Italy. They studied chronic paranoid schizophrenics and patients with neurotic-depressive reactions in addition to normals. It is noteworthy that their schizophrenic patients had not had drugs for at least one month and all subjects were male. Floris *et al.* (1968) presented curves showing the recovery functions of different latency measures in their three groups, but indicated that the differences between groups were not of particular significance. One latency difference to which they drew attention was a rise in R_2 latency between 100 and 180 msec in the normals, which was not observed in the schizophrenics. The main positive datum was in the amplitude recovery of negative peak 3, the same component studied by Speck *et al.* Figure 126 compares the mean recovery curves obtained in the three groups. The major differences occurred in the time intervals between 100 and 180 msec. Recovery during these intervals was significantly less in the patients than in the normals. The neurotic depressives show a slight facilitation around 70 msec, earlier than the normals, but after 100 msec their curve is not particularly different from that of the schizophrenics. More recently, Floris *et al.* (1969) reported additional data on visual recovery functions in patients with psychotic depressions and in epileptics. Recovery in psychotic depression was not different from that of normals. However, epileptics showed greater than normal recovery.

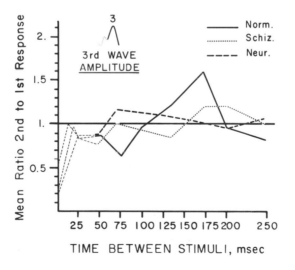

Fig. 126. Mean visual evoked response recovery functions for peak 3 in normal, schizophrenic, and neurotic subjects ($N = 15$ each). The peak of facilitation at about 170 msec, present in normals, is absent in the patient groups. (Reprinted with permission from Floris *et al.*, 1968.)

Comment

It will be noted that, although all three visual recovery function studies described here yielded positive results, the details of the findings significantly related to psychopathology differed from one study to another. Our own data showed only that latency recovery was slower than normal in the patients. In the study of Speck *et al.* amplitude recovery was reduced in patients, but particularly during an early phase of the recovery cycle, namely, the 35 msec interstimulus interval. In the Floris *et al.* studies, the main difference occurred in the later phase of the amplitude recovery curve (after 100 msec). These variations in findings may be due to a number of methodological differences between studies. Whereas Speck *et al.* used a flash stimulus, similar to that employed by us, Floris *et al.* employed a specially built photic stimulator; the flash source was placed 20 inches away from the eye, by Floris *et al.,* and 6 inches from the eye by Speck *et al.* and ourselves. The number of interstimulus intervals also differed considerably between studies, and the lead placements were not identical. The patients of Floris *et al.* were free of medication for a longer period than the patients of the other investigators. The pupils were mobile in all of the studies. Although the variations in method could have resulted in the different findings, it is noteworthy that all studies lead to the same general conclusion: that recovery is reduced in the presence of psychopathology. Furthermore, Heninger and Speck's (1966*b*) findings indicate that when psychopathology is diminished, visual recovery tends to shift toward normal values. The available results with visual recovery functions thus tend to confirm those obtained in the somatosensory modality. Not only is recovery found to be reduced in psychiatric patients, but there appears to be absence of diagnostic specificity.

CONTINGENT NEGATIVE VARIATION

The greatest volume of work on CNV reported in psychiatric patients comes from the Burden Institute laboratories of Grey Walter. Walter (1966) noted that in disturbed or "autistic" children, the CNV is absent altogether or appears only occasionally. He also reported that the CNV develops slowly and irregularly in patients with chronic anxiety states and rarely reaches the average amplitude of normal subjects. Furthermore, the introduction of three or four unreinforced trials (equivocation) was often enough to extinguish whatever CNV may have developed in anxious patients, and later reinforcement did not restore the response. Suppression of CNV in anxious patients was also readily produced by distraction.

In patients with compulsive-obsessional neuroses, Walter observed that the CNV, instead of declining to the base line as the button press was completed, either persisted or increased again to its original amplitude.

In addition, when unreinforced trials were introduced, CNV was more persistent in the compulsives. The most strikingly deviant CNV findings in association with psychopathology were obtained by Walter in patients considered to be psychopathic disturbances of antisocial type. Such subjects appeared to be incapable of producing more than a trace of CNV. In patients with schizoid features, Walter noted greater variability from trial to trial than normals, even though the CNV average did not seem to differ. Walter also presented some case reports in which patients who responded successfully to treatment developed a CNV, although it had been absent prior to therapy.

McCallum and Walter (1968) compared 40 chronic neurotic patients with 40 nonpatient control subjects, using a CNV paradigm in which the stimuli were paired click and flicker. Distraction was then introduced in the form of an irregular tone stimulus presented between trials. The patient group with high anxiety showed a significantly smaller basic CNV than the nonpatients; the mean of the patients was 16μv and that of the nonpatients 20μv. When distraction was introduced, the mean CNV of the anxiety patients fell to 4μv, whereas that of the nonpatients was reduced by only 6μv to 14μv (Figure 83).

Walter (1970) has reported additional data comparing CNV in various clinical groups, which appear to confirm and extend his earlier findings. Comparing various patient groups with age matched normal controls, Walter and his coworkers found a significant difference in basic CNV amplitude between the nonpatients and all patient groups. The psychopathic group continued to show virtually no CNV. The anxious patients also showed small CNV amplitude and marked defects of equivocation and distraction. Schizophrenic patients showed small CNV, greater effect of distraction, and greater extinction of the response. The mean CNV amplitude was about the same in the anxious patients and the schizophrenics, but variability was greater in the latter. The schizophrenics also showed a slow-negative wave after the response evoked by the imperative stimulus, which appeared as a continuation or second hump of the CNV in the acquisition trials. When CNV was reduced by distraction, the deflection following the response to the imperative stimulus remained.

Small and Small (1969) studied CNV in patients with affective psychoses in both the manic and depressive phases. After taking many precautions to reduce the influence of artifacts, particularly those introduced by the EOG, they felt satisfied in concluding that their data showed reduced CNV in patients with both phases of affective psychosis.

Dongier and Bostem (1967) failed to find CNV amplitude differences between normals and neurotics, but they did not employ a condition of distraction. Timsit et al. (1969) did obtain psychiatrically significant findings with the CNV method by focusing on the duration of negativity following the response to the imperative stimulus. In essence, their results agree with the observations of Walter (1970). Comparing 45 normals, 70 neurotics, and 45 psychotics, they found the duration of negativity after the imperative stimulus to be less than 1.5 sec in 91% of the normals, 66% of the neurotics, and only 7% of the psychotics. They also divided their neurotic group into hysterics and obsessionals and found that the duration of the negative deflection was shorter in the hysterics. Within their psychotic group, Timsit *et al.* found an intermediate duration (1.5 to 3.5 sec) in 8 manic depressives and a prolonged duration (more than 3.5 sec) in 13 out of 19 acute schizophrenics. The duration was also prolonged in 12 of 18 chronic schizophrenics. Timsit *et al.* also retested 16 psychotics after an interval of more than 15 days and found that 13 of them had a prolonged negativity of duration identical to that found in the initial test.

Although we have emphasized the major practical difficulties in obtaining valid CNV recordings in psychiatric patients, because of contamination by the EOG artifact (Straumanis *et al.*, 1969*a*), there seems to be little doubt that the CNV method has yielded results of psychiatric interest and will probably continue to do so. The extent to which the positive findings described above may reflect extracerebral events cannot be fully ascertained from the published reports. However, both Walter and his coworkers and Small and Small have taken special precautions to reduce such artifact and it would be surprising if a significant proportion of their findings did not reflect genuine phenomena.

COMMENT

The material presented in this chapter indicates that a fairly large number of significant evoked-response deviations from normal have been found in psychopathological states. With sensory responses proper, the most consistent findings have emerged with use of procedures that, to some extent, bypass the variations introduced by neuroanatomical givens, such as scalp thickness and brain position, by focusing on the interrelationships between responses. The measures of variability and of recovery function essentially employ this strategy and most of the studies that have utilized such measures have shown significant findings in psychiatric states. The CNV has also yielded a number of positive results, consisting mainly of reduced CNV amplitude and greater vulnerability to extraneous stimulation in psychiatric patients of various kinds. In addition, there have been several demonstrations that successful therapy of psychiatric illness is associated with normalization of initially deviant evoked potential measurements.

Although the evidence accumulated to date indicates that evoked-potential measurements have a promising future in psychiatric research, it should be emphasized that the surface has barely been scratched, and that the promise will probably take much time and labor to fulfill. One area of disappointment is in the relative lack of diagnostic specificity of those evoked-response indicators that do differ significantly with psychopathology. This is particularly so for recovery function measurements. The variability studies, especially those of Callaway and his group, suggest that variability may be more specifically disturbed in the schizophrenias, but more evidence is required. CNV is generally reduced in psychiatric states with the possible exception of obsessive-compulsive neuroses; Walter (1970) suggests that various patterns of CNV characteristics may be discriminative, but the observations upon which these suggestions are based await confirmation. The relative absence of diagnostic specificity need not, however, be discouraging. The exact psychological dimensions that are correlated with evoked-potential events remain to be defined. As indicated in the previous chapter, attempts to do so with various personality tests have not been rewarding so far, but it would be surprising, indeed, if physiological events were organized in exact correspondance to our psychological and psychiatric concepts. It seems clear that successful future developments await parallel improvements, in methodology and insight, at both the psychological and neurophysiological levels.

Perhaps the most important possible outcome of evoked-potential studies of psychiatric states resides in the focus they may give to animal studies. An understanding of the mechanisms that may give rise to increased evoked-response variability and reduced recovery could provide some lead as to the pathophysiological mechanisms involved in psychopathology. We have conducted a few such studies in our laboratory, following leads obtained in our clinical research. Among the more interesting findings was one showing that, in acute cat preparations, electrical stimulation of the mesencephalic reticular formation, at a rapid rate, markedly augmented early recovery of the cortical response to peripheral nerve stimulation. Figure 127 shows some of the results (Schwartz and Shagass, 1963).

In a more recent study conducted in chronically implanted cats (Shagass and Ando, 1970), the septal region was stimulated at both fast and slow rates electrically and the mesencephalic reticular formation (MRF) was stimulated at a fast rate. Responses were recorded from three cortical areas: primary somatosensory, motor, and associative. The results for all areas were generally the same. Stimulation of the septal region or MRF at 300/sec for 2 sec reduced the amplitude of R_1 and did not change the amplitude of R_2, so that the recovery ratio was significantly increased. The effect lasted for 15 to 30 min. Figure 128 shows the results for the somatosensory response. In contrast, slow stimulation of the septal region (5/sec for 10 sec) increased the amplitude of R_1 and decreased that of R_2 so that recovery ratio was reduced (Figure 128). It thus appeared that a mechanism that could produce the recovery function deviations found in psychiatric disorders would be reduction or slowing of the activity of the septal area.

The results of animal experiments, such as the foregoing, do not prove that similar mechanisms operate in mental illness, but they delineate

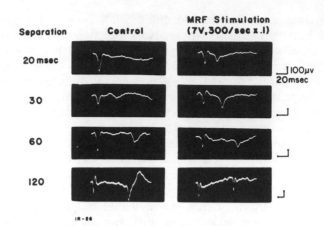

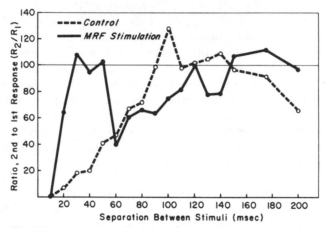

Fig. 127. Recovery to radial nerve stimuli at varying interstimulus intervals with and without prior stimulation to mesencephalic reticular formation (MRF). Acute cat preparation, immobilized with Flexadil, artificially respirated. The second response is markedly augmented at intervals of 20 to 50 msec by MRF stimulation. Downward deflection indicates positivity. (Reprinted with permission from Schwartz and Shagass, 1963a.)

relevant possibilities. Further experimental manipulation, perhaps by drugs, could then lead to some testable hypotheses concerning the possible role played by such mechanisms in the mental disorders themselves.

It may be worthwhile in the present context to draw attention to two major points of view that underlie the electrophysiological approach to studies of mental disorder. One point of view may be called pathophysiological and the other psychophysiological. Our own view is a pathophysiological one, in the sense that we expect our evoked-response data to reflect basic underlying disturbances of CNS activity, perhaps at the neuronal level in terms of altered electrolyte balance, or in altered relationships between excitability at various levels of the brain. In looking at evoked-response events, we do not necessarily expect to find any one-to-one correspondence between momentary electrical phenomena and behavior. In contrast, the psychophysiological view does expect to find clear correlations between momentary behavioral events and the electrical recordings. From the psychophysiological point of view, electrical phenomena ought to deviate from normal because the behavior of the subjects deviates from normal. However, from this point of view also, the deviations are normal in themselves since the electrophysiology is closely tied to the behavior being observed.

Both the pathophysiological and the psychophysiological points of view have validity and there is no clear indication that one will be more or less productive than the other in this area of study. It seems of value, however, to make the issue clear, since the point of view of the investigator does have some effect on the methodology employed in evoked-potential studies of psychiatric disorder. For example, in our own studies, we have emphasized the early evoked-response events because these are least likely to be susceptible to changes as a consequence of transitory psychological

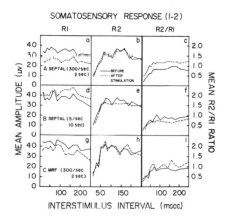

Fig. 128. Recovery measurements made from electrode chronically implanted over primary somatosensory cortex of cat. *A,* before and after septal stimulation at 300/sec for 2 sec; *B,* before and after septal stimulation at 5/sec for 10 sec; *C,* before and after MRF stimulation at 300/sec for 2 sec. Note that rapid stimulation to septum or to MRF reduces R_1, whereas slow stimulation increases R_1. R_2 is relatively unchanged by stimulation, so that R_2/R_1 ratio is increased by fast and diminished by slow stimulation. (Reprinted with permission from Shagass and Ando, 1970.)

reactions (Chapter 6). Our recent study, comparing somatosensory-recovery functions while subjects were viewing television and while they had their eyes shut, was conducted to determine whether differences in attentiveness influenced recovery measurements. The negative results were taken as reassuring, in the sense that they do not support the hypothesis that patient–control differences result from group differences in state of attentiveness (Shagass *et al.*, 1970). In contrast, Callaway and his coworkers deliberately set out to test a hypothesis concerning deviant attentive behavior in schizophrenia and have focused their measurements upon later evoked-response events. Similarly, Grey Walter, in his CNV work, is using a technique that depends upon manipulations of attentiveness. The apparently consistent CNV findings in different psychopathological states must, therefore, be interpreted as evidence of altered states of attentiveness and susceptibility to distraction. In other words, it is as easy to conclude that the physiological responses are different because the behavior of mentally ill people is unusual as it is to conclude that the patient is behaving in a deviant manner because his brain is not functioning properly. The kind of information-processing interpretation given to evoked-response variability by such workers as Lifshitz falls within the pathophysiological framework, since it is assumed that the patient behaves peculiarly becuase his brain cannot process information properly.

Chapter 10

Effects of Pharmacologic Agents

The study of psychoactive drugs offers one of the more interesting applications of evoked-response methods to psychiatry. Evoked-potential changes that accompany behavioral effects of drugs could provide clues concerning neural mechanisms that mediate both drug action and the associated behavior. However, to establish the relevant relationships, important methodological difficulties must be resolved (Shagass, 1968c). The drug effects that are psychiatrically interesting are those involved in reversal of psychopathology (e.g., the antipsychotic effect of phenothiazines). The electrophysiological correlates of such specific therapeutic effects must be distinguished from those accompanying other behavioral changes, such as drowsiness. They must also be separated from electrical effects that may be regarded as "tissue responses" without known behavioral correlates. Examples of "tissue responses" would be the EEG fast waves produced by barbiturates (Brazier and Finesinger, 1945) and probably the increased negativity of the somatosensory peak at about 35 msec (described below).

Changes related to alertness are of particular importance in evaluating evoked-response effects of psychoactive agents since so many drugs alter consciousness. Figure 129 illustrates the problem. It shows three sample tracings taken from a narcotics addict at different stages of withdrawal from methadone. Tracing *A* was taken when the patient was receiving 80 mg of methadone and 80 mg of chlordiazepoxide per day. Tracing *B* was taken when he was receiving a reduced dose of chlordiazepoxide and no methadone, but was showing withdrawal symptoms; the tracing resembles that seen in drowsiness, with a large secondary component and prolonged latencies. Tracing *C* was recorded when the patient had been successfully withdrawn from both drugs and his sensorium was clear; it resembles the record taken when he was receiving maximum drug dosage. There were, thus, marked changes when the state of alertness was altered, but the direct effect of drugs on the somatosensory response was minimal.

Since the subject in Figure 129 had a clear disturbance in sensorium, it was not difficult to attribute evoked-response changes to his altered state of alertness, but the minor variations in alertness produced by many drugs would be more difficult to evaluate. The evoked-response changes associated with them would be correspondingly hard to disentangle from those more directly related to modification of psychopathology.

Some of our findings illustrate another methodological problem of importance to psychopharmacology; this concerns the psychiatric status of subjects to whom drugs are administered. Although most workers are aware of possible species differences in drug effects, it is not so widely recognized that effects may differ quite markedly between healthy and sick people. In the preceding chapter, Figure 98 showed that the somatosensory intensity–response curve was reduced by imipramine therapy in a psychotically-depressed patient. Since the psychotic-depressed patients in our study generally had high-amplitude responses, the effect of the imipramine could be regarded as normalizing. In contrast, the effect of imipramine in therapeutic dosage administered to nonpatient volunteers was quite the reverse (Shagass *et al.,* 1962); the intensity–response curve was elevated after treatment (Figure 130). The results in these nonpatients were in accord with observations on imipramine reported in the rabbit by Van Meter *et al.* (1959). Consequently, although the drug findings in presumably normal man and rabbit agree, the results could not be generalized to the depressed

A. 10-25-63
BEFORE WITHDRAWAL-Methadone 80mg/day
Librium 80mg/day

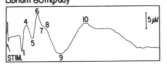

B. 11-5-63
METHADONE STOPPED 9DAYS-Librium 80mg/day
To 11-4 Now Reduced to 40mg/day-Speech
Slurred, Irritable and Drowsy

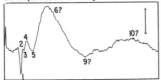

C. 11-12-63
LIBRIUM DISCONTINUED 24HR-Patient Seems
Cheerful, Well Coordinated

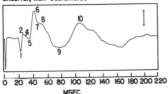

0 20 40 60 80 100 120 140 160 180 200 220
MSEC.

Fig. 129. Somatosensory responses during varying states of drug intake and alertness in a methadone addict. The responses in *A* and *C* appear similar even though drug intake was quite different, whereas response in *B* is markedly different in association with altered state of sensorium. (Reprinted with permission from Shagass, 1968*c.*)

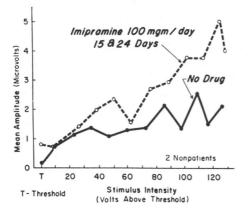

Fig. 130. Mean amplitude–stimulus intensity function for initial negative–positive somatosensory component in two nonpatient subjects before and after chronic intake of imipramine. (Reprinted with permission from Shagass *et al.*, 1962.)

patient. With respect to the effect of imipramine on somatosensory responses, the nonpatient and the depressed patient seem to be representatives of different species.

PREANESTHETICS AND ANESTHETICS

Preanesthetics

An extensive series of observations on the effects of preanesthetics has been reported by Domino and his coworkers, who studied visually evoked responses (Domino *et al.*, 1963; Corssen and Domino, 1964; Domino and Corssen, 1964; Domino, 1967). The dosages used were those commonly employed for preanesthetic purposes. Drugs studied were: muscarinic cholinergic blocking agents (scopolamine hydrobromide, methscopolamine bromide, atropine sulphate, and l-hyoscyamine hydrobromide); tranquilizers (chlorpromazine hydrochloride, promethazine hydrochloride); analgesics (morphine sulphate); sedatives (secobarbital sodium, chlordiazepoxide); psychomotor stimulants (*d*-amphetamine sulphate). Of all these agents, scopolamine hydrobromide was most effective in altering the visual-evoked response. It reduced amplitude and increased the latency of early components, while at the same time it appeared to cause augmentation of amplitude and rhythmic activity in the later portions of the response. In contrast, methscopolamine, which is known to penetrate the blood-brain barrier with difficulty, had no significant effect. The sedative and tranquilizing agents tended to alter the response in a way similar to that found with natural sleep. Domino (1967) concluded that the effects produced by preanesthetic agents were related more to the physiologic state of the patient than to the drug given. He suggested that a drug that promotes alpha rhythm in the EEG will enhance the visual evoked response, including the after-rhythm, whereas drugs that induce sleep tend to produce changes typical of natural sleep.

Inhalation Anesthetics

Domino *et al.* (1963) found that nitrous oxide was relatively ineffective in depressing the visual responses and occasionally enhanced the "primary" complex. More systematic effects were shown by Lader and Norris (1968) with click-evoked responses; these were reduced in amplitude by nitrous oxide and the degree of reduction was a function of the concentration of the anesthetic agent, even though their doses were actually subanesthetic. Bennett *et al.* (1969) showed that hyperbaric nitrogen and oxygen both depressed the amplitude of the click-evoked response. They concluded that the compressed-air narcosis experienced by divers is due mainly to nitrogen, and that the auditory response provides a reliable and reproducible measure of narcosis.

Diethyl ether and cyclopropane, in amounts sufficient to produce clinical anesthesia, were especially effective in depressing the visual evoked response (Domino *et al.*, 1963). Other inhalation anesthetics produced different effects; halothane and methoxyflurane in intermediate doses enhanced the secondary waves, especially the positive component occurring about 120 msec after the stimulus. Domino *et al.* indicated that all anesthetics studied appeared to enhance the late positive wave at critical intermediate depths of surgical anesthesia.

Barbiturates

Cigánek (1961*a*) reported that the visual-evoked response changed with thiopental (Pentothal) narcosis. The earliest components were relatively unchanged, except for prolongation of latency, but the amplitude of his negative wave V (average peak at 114 msec) was markedly augmented when the EEG showed slow activity. Wave VI was also augmented in deep narcosis, and the after-rhythm was abolished. Cigánek concluded that the effects of sleep were similar, whether natural or induced by barbiturate or chlorpromazine. Domino *et al.* (1963) and Domino (1967) found visual response changes induced by thiamylal sodium (Surital®) to be dose-dependent. The amplitude of the initial waves was enhanced in light anesthesia, whereas the entire response was flattened in deep anesthesia. These workers conducted parallel studies in monkeys which indicated that the depressant effects of thiamylal were directly on the visual cortex or on some afferent system impinging upon it, since they were not observed in recordings from the lateral geniculate radiations. Bergamasco (1967) failed to observe modifications of the earlier components of the visual evoked response during light barbiturate narcosis, but noted that they tended to disappear with deep anesthesia. The later components were present in both light and deep anesthesia.

Rosner and his coworkers studied the effects of thiopental on somato-sensory evoked responses (Allison *et al.*, 1962; Abrahamian *et al.*, 1963;

Rosner *et al.*, 1963). Their subjects were premedicated with secobarbital 100–160 mg and atropine 0.4–0.6 mg, and the premedication produced some changes in late portions of the response. Thiopental, administered intravenously in successive doses of 10 to 100 mg each, abolished later portions of the response and then slowed and reduced earlier deflections. At lower doses of thiopental, the initial or primary portion of the somatosensory response was not affected, or even augmented. However, a marked negativity developed, or was unmasked, between the first and second positive components, and the second positive component occurred much later in time and was markedly reduced in amplitude at anesthetic dose levels. Rosner *et al.* (1963) showed comparative tracings in monkey and man which demonstrated remarkable similarity of the effects of thiopental in the two species. They concluded that the results with thiopental anesthesia were in accord with those obtained in other experiments in indicating that the first rapid triphasic complex of the human contralateral somatosensory response represents activity in the thalamocortical afferent axones. The negative wave, which appears under anesthesia and in recovery function measurements with brief interstimulus intervals, was thought to be present normally, but usually masked by the second slow positivity.

The changes in early somatosensory response components produced by thiopental in the studies of Rosner *et al.* are similar to those observed by us following intravenous amobarbital in sedative doses (Shagass *et al.*, 1962a). However, as illustrated by Figure 131, this effect of barbiturates does not

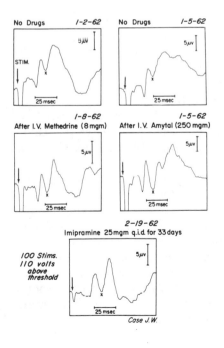

Fig. 131. Effects of different drugs on somatosensory response of one patient. The deflection marked by X is made more negative after single injections of methedrine and amytal and after prolonged intake of imipramine. (Reprinted with permission from Shagass and Schwarz, 1964b.)

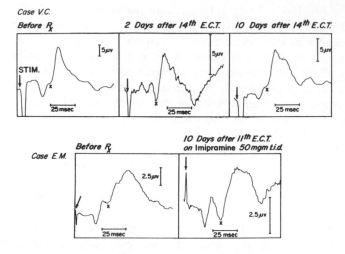

Fig. 132. Augmentation of second negative deflection *(X)* by ECT in case VC and by combination of ECT and imipramine in case EM. Note that change disappears 10 days after last ECT in case VC, but persists in EM, who continued to receive imipramine. (Reprinted with permission from Shagass and Schwartz, 1964*b*.)

appear to be specific, since we observed it in the same subject with acute doses of both amobarbital and amphetamine, and also after the patient had been receiving imipramine in therapeutic doses for a time. We have also observed the same effect with phenothiazines, LSD, and electroconvulsive therapy (ECT). Figure 132 shows the changes with ECT; it also shows that, whereas the ECT effect disappeared after 10 days, it was present when imipramine therapy was continued after ECT. The greater negativity of peak 5 in tracings *A* and *B* than in tracing *C* of Figure 129 shows that chlordiazepoxide, and possibly methadone, also probably produce this nonspecific effect.

Ikuta (1966) administered thiopental to a group of schizophrenic patients and studied the effects of different doses on somatosensory responses. He found that latency of the main positive and negative peaks increased proportional to dosage. The latency of later peaks increased more than those of earlier ones, and he observed latency increases even when sleep did not occur. He found that changes in amplitude were inconsistent; they differed markedly between subjects and with different doses. Our own study of the effect of amobarbital on the somatosensory intensity–response curve failed to reveal consistent changes in the amplitude of the initial negative–positive component (Shagass *et al.*, 1962).

Steroids

The steroid anesthetic, Viadril, changes visual and somatosensory responses. Hirsh *et al.* (1961), in an important study that showed the similarity between evoked responses recorded from the scalp and those recorded directly from the brain at surgery, observed that Viadril increased latency. They did not find similar effects with local anesthesia. The effects of Viadril, a steroid with anesthetic properties, appear to differ from those of steroids with other pharmacologic effects, since Ojemann and Henkin (1967) found that carbohydrate-active steroid treatment of patients with adrenal cortical insufficiency decreased latency of visually evoked responses. The changes were most pronounced in the later waves. Latency decreases were not observed after treatment with sodium-potassium active steroids. Ojemann and Henkin suggest that the latency changes observed by them may be related to alterations in concentration of brain steroids, either through a direct action on brain tissue or on ion balance across cell membranes. It would be of interest to know whether similar latency changes are observed in normal subjects with chronic steroid administration. Kopell *et al.* (1970*a*) found that a single dose of cortisol increased the latency of an attention-related visual-response late wave in normals and decreased its amplitude.

PSYCHOACTIVE AGENTS

Sedatives

Although Corssen and Domino (1964) found no consistent effect of chlordiazepoxide on the visually evoked response, Bergamasco (1966, 1967) and Poiré *et al.* (1967) showed that the related compound, diazepam (Valium®), reduced amplitude. Poiré *et al.* noted, in addition, that the after-rhythm was abolished. Ebe *et al.* (1969) also found that diazepam reduced the amplitude of both somatosensory and visual responses. Dolce and Kaemmerer (1967) found that another compound of the series, oxazepam (Serax®), produced a reduction of amplitude and slight latency increases in the response to flash. Prieto *et al.* (1965) observed no effects of meprobamate (800–1200 mg) on visually evoked responses. Secobarbital has been noted to increase the amplitude of visually evoked responses in association with sleep (Corssen and Domino, 1964).

Alcohol

Gross *et al.* (1966) showed that alcohol reduced auditory response amplitude. Lewis *et al.* (1970) found that 3 oz of alcohol reduced amplitudes of late components in both visual and somatosensory responses recorded from central areas. Early components were relatively unchanged. A smaller dose of alcohol (1 oz) had little effect. Responses recorded from occipital leads were not affected by alcohol. Lewis *et al.* interpreted their results as suggesting that alcohol exerts a depressant action first on subcortical areas, such as reticular formation, rather than at cortex, as long believed.

Excitants

Bergamasco (1966) found that pentamethylenetetrazol (Metrazol) caused a slight increase in amplitude of visual evoked responses and shortened the visual cortical recovery cycle. His results are in agreement with the early ones of Gastaut *et al.* (1951*a*). Crighel and Ciurea (1966) studied the effects of *d*-amphetamine sulphate on the visual evoked response; they noted that, with doses that do not modify the EEG, amplitude and variability were increased. However, Domino and Corssen (1964) observed no significant effects with *d*-amphetamine. We found no consistent effects of intravenous methamphetamine on the initial component of the somatosensory response, although this agent, in common with others, augmented the second negative component (Shagass *et al.*, 1962). Bishop *et al.* (1969) studied the effects of a combination of dextroamphetamine and amobarbital on visual responses and found shortening of latency, suggesting that the amphetamine effect was dominant over that of the sedative.

Major Tranquilizers

Cigánek (1959) found that chlorpromazine had no effect on the early visual-response components. In our own work, we have been particularly interested in drug effects that may persist for some time after medication is withdrawn, because it is difficult to obtain patients for study who have been withdrawn from medication for more than a few days. To estimate residual drug effects, we have repeatedly compared results in patients who had received some phenothiazines from 1 to 7 days before testing with those of patients who had not received drugs for 8 days or more. These comparisons have never demonstrated significant differences between groups in the amplitudes and latencies of early components of visual and somatosensory responses, or of their recovery functions. On the other hand, several workers have reported that phenothiazines do modify evoked responses.

Heninger and Speck (1966) showed marked visual response effects of phenothiazine medication in schizophrenic patients. That some of these effects could have been due to drowsiness induced by drugs is suggested by the increased amplitudes. This possibility is supported by the fact that the records obtained by Heninger and Speck after 30 days of treatment were more like those obtained before treatment than those displayed a short time after therapy had been started. Corssen and Domino (1964) observed some changes with chlorpromazine and promethazine, which they attributed to alterations in alertness. Helmchen and Kunkel (1964) found that schizophrenic patients treated for several weeks with perazine had EEG slow waves and displayed even slower visual-response after-rhythms.

Prieto *et al.* (1965) reported some visual-response changes with an unspecified dose of chlorpromazine. In a preliminary report, Jones *et al.* (1969) noted that somatosensory responses were modified in various ways by a series of agents which included: perphenazine, fluphenazine, thiothixene, haloperidol, and dibenzoxephine. Both increases and decreases in amplitude and latency, particularly of a positive component at about 65 msec, were observed. The effects were related to dosage and duration of treatment. Recently, Saletu *et al.* (1970) have reported that thiothixene decreased somatosensory evoked-response amplitude. An interesting aspect of the the Saletu findings is that changes were observed only in patients who showed a favorable clinical response.

There is some evidence that those evoked response indicators that have been found to be deviant in psychosis tend to normalize when phenothiazine therapy has been effective. Jones *et al.* (1965) reported that the two-tone similarity score became higher after effective treatment. In an early study, we found some tendency for normalization of recovery function after effective phenothiazine therapy (Shagass *et al.,* 1962). Heninger and Speck (1966) found a shift toward normalization of visual recovery function. Preliminary results with our modified recovery-function procedure in patients studied serially, before and during treatment with phenothiazines, showed no consistent effects for the group as a whole. However, division of the patients into two groups according to clinical response (good and poor) revealed some interesting trends (Figure 133). The patients who responded favorably to phenothiazines showed higher-amplitude somatosensory responses before therapy than the poor responders; they also demonstrated relatively steep R_2 and R_{10} gradients as a function of the intensity of the conditioning stimulus. In contrast, the poor responders had relatively flat pretreatment R_2 and R_{10} curves, which resembled those of the psychotic drug abusers in Figure 124. While receiving phenothiazines, the good clinical responders showed reduced amplitude of R_1 and a relative increase in R_2 and R_{10}, whereas the poor responders showed very little change in any of their curves. Our results with the modified recovery function procedure seem to agree with those of Saletu *et al.* (1970), in that changes with drugs were found only in clinical responders. Also, it appears that the response of schizophrenic patients to phenothiazines may, to some extent, be related to their pretreatment evoked-response characteristics.

One interesting aspect of the results in our favorable clinical response group is that the R_2 changes with treatment resemble those obtained following septal or reticular electrical stimulation at a fast rate (Shagass and Ando, 1970). Since such stimulation presumably increases septal and reticular activity, this suggests that the antipsychotic effect of phenothiazines may involve heightening of activity in these subcortical structures. This is a somewhat unexpected implication of the findings, as most workers believe that the reverse effect may occur. However, the reduced cortical recovery found in psychiatric patients appears to be paralleled in animals when

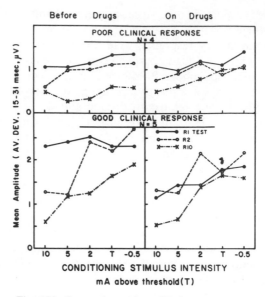

Fig. 133. Comparison of modified somatosensory recovery function results in 4 schizophrenic patients who responded poorly to phenothiazines and in 5 patients who responded well clinically. Note relative absence of R_2 and R_{10} gradients in patients with poor clinical response and continued absence of gradients while on drug. Patients with good clinical response show relative increase of R_2 and R_{10} (compared to change in R_1 test) while on drugs.

septal activity is slowed by low-frequency electrical stimulation. Consequently, one would expect both increased cortical recovery and clinical normalization to accompany increased septal activity. The poor clinical response to phenothiazines in patients with evoked-response characteristics that suggest an already highly activated state of septal or reticular activity also appears to be consistent with the idea that phenothiazines may increase septal activity. These patients would be at the other extreme of those who respond. The implication would be that the desirable level of septal or reticular activity is somewhere in between that of the responders and the nonresponders.

Antidepressants

We found that imipramine and tranylcypromine (Parnate®) returned the somatosensory recovery function toward normal and diminished amplitude in patients with psychotic depressive reactions (Shagass and Schwartz, 1962a). Imipramine also augmented the second negative deflection of the somatosensory response (Figure 132).

Lithium

The therapeutic effectiveness of lithium in the manic phase of manic-depressive illness is now well documented. Heninger (1969) found that lithium increased the amplitude of the initial negative–positive component of the somatosensory response. We have confirmed his results in our laboratory with data obtained in the context of our modified recovery-function procedure. Figure 134 shows tracings obtained before treatment and while the patient was receiving lithium; in every case, the response after the drug was of greater amplitude. For our sample of six patients, an analysis of variance showed a highly significant treatment effect. The mean $R_1 T$, R_2, and R_{10} curves for the group are shown in Figure 135. Before lithium, the mean curves varied little in relation to conditioning stimulus intensity and were relatively flat. The lithium curves all had steep gradients. The changes in the gradients with lithium were statistically significant. The intensity–response curves, based on amplitude of the responses to the conditioning stimuli alone, were also different with lithium; mean amplitude was greater and the slope from the weakest to the strongest stimulus was steeper.

Gartside *et al.* (1966) studied somatosensory recovery functions in nine nonpsychotic volunteers who took lithium carbonate for several weeks. Figure 136 illustrates their findings. The recovery function shifted from the pattern found in normals to that found in psychiatric illness. They failed to observe changes in amplitude of the response to single stimuli; since the dose used, 900 mg/day, was smaller than that used by Heninger or ourselves, the discrepant findings may involve variation in dosage. Gartside *et al.*

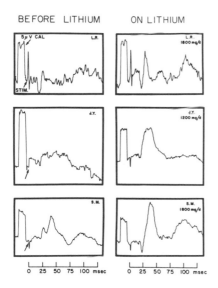

Fig. 134. Effects of lithium on somatosensory evoked response in three patients with affective psychoses.

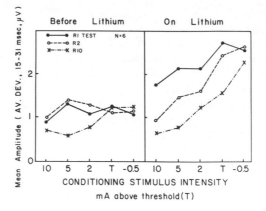

Fig. 135. Effects of lithium on modified somatosensory recovery test. Mean of 6 patients with affective disorders. Note markedly increased R_2 and R_{10} gradients.

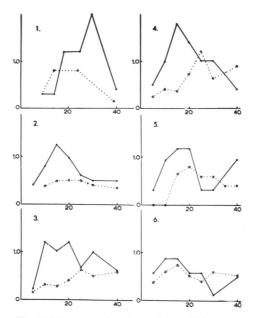

Fig. 136. Recovery functions of the initial somatosensory response component before and after lithium carbonate administration. Ordinate is R_2/R_1 ratio; abscissa, interstimulus intervals. Six different subjects. Before lithium, recovery curves are normal; with drug, the early phase of the recovery function is suppressed. (Reprinted with permission from Gartside *et al.*, 1966.)

had hoped that shifts of the recovery function would be associated with clinical signs of depression, but this did not prove to be the case. If our values for the 10 ma above threshold conditioning stimulus intensity, shown in Figure 135, are compared with those obtained by Gartside *et al.* (Figure 136), it will be seen that the results are very similar. The recovery ratio (R_2/R_1T) is above unity before lithium and well below unity in the drug data for the subjects of both studies.

Small *et al.* (1970) found that lithium prolonged peak latencies of both auditory and visual responses. The changes were stable for the visual responses and transitory for the auditory. They also observed lithium effects on the CNV; lithium increased the CNV (increased negativity).

The findings so far obtained with lithium suggest that the drug increases cortical responsiveness to single peripheral nerve shocks, and brings about a greater dynamic range of responsiveness to varied stimulus conditions. With intense stimuli, it reduces recovery; this may possibly be interpreted as a mobilization of inhibitory activity, which does not occur with weak conditioning stimuli. The marked change in the initial components of the response is also of great interest since the effect must occur at one, two, or all of only three possibly synapses. It should be possible to determine which synapses are involved in animal experiments.

THYROID FUNCTION

Thyroid hormone has been extensively used in the past in attempts to modify psychiatric states. A notable example is the periodic catatonia of Gjessing. Mental changes associated with abnormal thyroid function have also received much study. In addition, it is possible that some psychoactive drugs exert their effects, in part, by modifying thyroid activity. Consequently, the fact that there are significant evoked-response differences in conjunction with variations in thyroid functioning is of considerable interest in the present context.

Nishitani and Kooi (1968) studied visual evoked responses in 45 patients with hypothyroidism and 45 matched normal subjects. They found that average amplitudes were smaller in the hypothyroid patients with one exception, namely, the amplitude of the second occipital positive wave was larger. Average latencies of all components examined were longer for the hypothyroid patients. When thyroid replacement therapy was instituted, the responses normalized and the differences between patients and controls disappeared.

Findings in the reverse direction were obtained by Short *et al.* (1968). They administered triiodothyronine to eight nonpatient control subjects for three days. The results are illustrated in Figure 137. Latency and amplitude of the waves from IV on were increased and the after-rhythm was potentiated while the patients were receiving the drug. Kopell *et al.* (1970*b*) observed

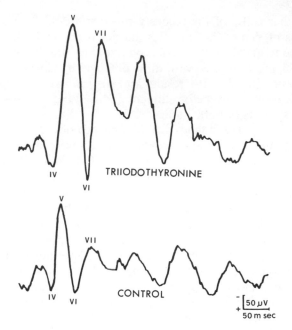

Fig. 137. Effect of 3 days administration of triiodothyronine. Nonpatient subject. Visual responses are the
average of 60 flashes administered at 1/sec. Note marked
increase of amplitude. (Reprinted with permission from
Short *et al.,* 1968.)

increased amplitude of a late wave related to attention when triiodothyronine was administered; however, the change was observed only with stimuli
that the subject ignored and not with those to which he attended.

It is noteworthy that the effects on the visual response of either increased or decreased thyroid functioning seem to occur mainly in the
secondary waves. The investigators whose work has been cited here have
speculated about the possible relationships to attentive activity. Short *et al.*
interpret the facilitation of the secondary component with triiodothyronine
to indicate a relative reduction in the level of ascending reticular activating
system function.

PSYCHOTOGENIC AGENTS

The evoked-response changes induced by the anticholinergic deliriant,
Ditran, have been described previously (Chapter 5) in relation to states
of impaired awareness. Sernyl, an agent with properties similar to Ditran,
was studied by Rodin and Luby (1965). They found that the effects of Sernyl

were variable from one subject to another, but that there was a tendency toward amplitude decrease in the visual evoked response and an increase of slow activity in the EEG.

Rynearson et al. (1968) studied the effects of psilocybin, an agent with clinical effects similar to LSD. The dose was 10 mg p.o. and a placebo comparison was made in the same nonpatient subjects. Evoked responses to flash were definitely changed in only four of 22 subjects; the changes involved reduction of afteractivity, as well as a variety of differences in early components. It was of interest that those changes occurred in relation to experience of visual symptoms.

The effects of lysergic acid diethylamide (LSD) have been investigated by several workers. In an early study, we found that there was no quantitative change in the initial somatosensory component amplitude (Shagass et al., 1962). In a dose of 1 µg/kg iv, LSD shared with other agents the tendency to augment the second negativity (Shagass and Schwartz, 1964a). There appeared to be no direct association between the behavioral effects of the drug and changes in the somatosensory response. Rodin and Luby (1966) studied the effects on visual responses of LSD (1 µg/kg). They found that the components occurring in the first 250 msec tended to decrease in amplitude; there was a slight decrease in latency in half of their eight subjects. After-rhythms were reduced in amplitude in six subjects and there was an associated reduction of EEG activity with frequency less than 24 c/sec. An important aspect of the Rodin and Luby study is that they administered a mydriatic before giving LSD and that they gave LSD by infusion. Furthermore, the maximum behavioral effects of LSD were noted after the EEG observations were terminated. Chapman and Walter (1965) reported, as did Rodin and Luby, that the most striking effect of LSD in doses of 1.4 to 2.3 ug/kg, p.o. was a reduction of the visual after-rhythms. Prieto et al. (1965) found that LSD, 100 mcg im, increased the amplitude of some components and reduced variability of the visual response.

Brown (1969) compared visually evoked responses obtained before and after LSD administration in subjects classified into two groups, those possessing vivid visual imagery and those with total absence of visual imagery ability. Various colors of light flash were used. Before LSD, the nonvisualizer subjects had had similar responses to virtually all colors but, following LSD, the early components were changed in configuration as a function of color, most clearly for red and green; changes were mainly in latency. The visualizer subjects showed significant variation with color before LSD and selective alterations in color specificity after LSD. The changes were more pronounced in the realm of amplitude. The visualizers also reported enhanced visual perception with LSD. Brown's results, which suggest differential reactivity to LSD of visual evoked responses as a function of visualizing ability, should be considered in relation to the EEG results that she had reported in the same subjects (Brown, 1968). LSD produced increased EEG activation in visualizers and reduced EEG activation in nonvisualizers. Visualizing ability would thus seem to provide one source of variation in electrophysiological responsiveness to LSD.

In our most recent study of the effects of LSD on evoked responses (Shagass, 1967*a*), we recorded from six areas simultaneously and observed effects on the after-rhythm of the visual response that differed from those reported by other investigators. Our results were consistent with those of others only in recordings from bipolar leads placed over the optimal visual area. However, the amplitude of after-rhythm was more often actually increased in areas peripheral to the "central" pair of leads (Figure 40). The frequency of after-rhythms was also increased, as was the general EEG frequency. It appeared that the area of the brain giving rise to synchronized after-rhythm was increased by LSD, so that there was little potential difference between the "central" electrodes; this led to the impression of reduced amplitude in that derivation. The earlier components of the visual-evoked response were also found to be somewhat reduced in this study, a result in accord with that of Rodin and Luby (1966).

We also recorded the effects of LSD on somatosensory responses to stimuli of 10 ma above threshold. Figure 138 illustrates the results. There was a reduction in amplitude and latency of the early components. The subjects of our study were patients with conduct disorders undergoing therapy with LSD (Shagass and Bittle, 1967). The data suggested that those individuals whose evoked responses were larger initially, and were more reduced in amplitude by LSD, later showed more behavioral improvement.

The electrophysiological effects of LSD resemble those found in alerted states. This is in contrast to the effects of deliriants, such as Ditran and Sernyl. The central effects of these psychotogenic agents are obviously different, as are the behavioral effects (compare Figures 40 and 72). The

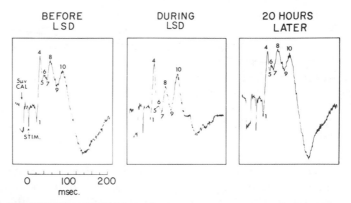

Fig. 138. Effect of LSD 2.5 mcg/kg, iv, on somatosensory evoked response. Effects are gone 20 hr later. (Reprinted with permission from Shagass, 1967*a*.)

patient receiving LSD rarely forgets the experience and, when he reports his thoughts while under the influence of the drug, he may describe very rapid mental activity and intense perceptions. The patient receiving Ditran is nearly always amnesic for the experience and his thoughts are uncoordinated and his mental activity is generally incoherent when he is under the influence of the drug. The electrophysiological differences between these agents seem to be in general agreement with the behavioral effects. EEG slowing and evoked-response changes in secondary components, that are suggestive of sleep, occur with Ditran, while EEG speeding and evoked-response secondary changes, suggestive of heightened alertness, are seen with LSD. It may be that the behavioral dimension that correlates best with electrophysiological phenomena is the perceptual–cognitive sphere designated clinically as "sensorium" (Shagass, 1968e).

COMMENT

Clearly, many evoked-response changes occur as a consequence of drug action. Two questions concerning these are of paramount importance to the psychiatric investigator. One is methodological: How long do patients have to be off drugs before their records can be considered representative of the undrugged state? There are indications that the main effects of drugs may not persist for longer than a few days at most. Nevertheless, in view of the EEG findings of Ulett et al. (1965), more definitive data are urgently needed. It may be that drug effects on evoked responses persist, as they do on the EEG, for many weeks. The other question concerns the drug changes that relate to reversal of psychopathology. In view of the difficulties involved in separating such effects from those related to level of alertness and "tissue responses," it can only be said that no definite answers are available, although some interesting data have emerged.

The following findings discussed in this chapter may perhaps be singled out as being of major interest:

(1) The response-depressant effects of diazepam and oxazepam, which suggest reduction of excitability. The antianxiety effects of these agents would then be a function of increased inhibition or decreased excitation.

(2) The major effects of anesthesia on late components; these suggest modification of activities involved in the processing and storage of sensory information.

(3) The contrasting effects of Ditran and LSD, pointing to sensorium as the psychological dimension that may relate best to electrophysiological phenomena.

(4) The effects of lithium carbonate, which suggest that this agent may exert some of its important actions on the synapses of the primary

afferent pathways. Also, the evidence that lithium tends to normalize the dynamic range of cortical responsiveness to sensory stimulation suggests that it improves the functioning of central mechanisms
regulating the balance between excitatory and inhibitory processes.

(5) The data suggesting that drug effectiveness depends upon the pre-
existing state of cortical responsiveness, e.g., the differences in
response to imipramine in psychotically depressed patients and
nonpatients, and the different clinical response to phenothiazines
of schizophrenics as a function of quantitative evoked response
patterns.

(6) The evidence indicating that at least the later components of
evoked responses are related to the state of endocrine function,
particularly in the spheres of thyroid and adrenal activity.

It seems reasonable to assert that when findings such as these are
elaborated, our understanding of brain–behavior relationships will be
enhanced.

Chapter 11

Conclusion

Evoked-potential recording is now a well established laboratory technique. It remains primarily a research method, although it has found some practical applications in clinical diagnosis, e.g., in audiometric testing of nonverbal subjects (Callaway, 1969). The use of the method in clinical research has not been without controversy. This is of the kind ordinarily generated between more "basically" and more "clinically" oriented investigators when a new biological method appears. Those with a "basic" orientation believe that application of the method to clinical problems can have little value until all parametric studies have been performed and the factors that require experimental control are known. Some would go further and say that clinical correlations will be meaningless until the underlying physiological mechanisms have been worked out. On the other hand, the clinically oriented investigator takes the position that a major investment in parametric studies and research on mechanisms can be justified only by evidence that the biological phenomenon is clinically relevant. There is justification for both points of view. The view that has guided our own work is obviously the clinical one, since most of our parametric and animal studies have followed, rather than preceded, the finding of clinical correlates.

From the clinical point of view, an evaluation of the achievements of evoked-response research in psychiatry brings a mixed reaction. There is no doubt that there are enough positive results already in hand to justify the assertion that evoked response methods will play an important role in biological psychiatry for a long time to come. Clearly, any method possesses psychiatric merit if it can yield objective measurements of brain activity that are correlated with state of consciousness, intelligence, personality, presence of psychopathology, and psychoactive drug effects. However, it must also be stated that, so far, the results which would be of the greatest interest to psychiatry are deficient in many respects. We

241

lack evoked-potential correlates of specific psychiatric diagnostic entities, syndromes, or symptoms. This lack does not yet represent a definitive negative result, but the balance of evidence suggests that evoked potentials probably will not generate measurements with a one-to-one correspondence to our currently accepted clinical concepts. While this lack of correspondence should hardly be surprising, it is disappointing. It is important, therefore, that the disappointment be kept in proper perspective.

At the present time, we are so bound by our clinical concepts, that we may be unable to perceive some useful new dimensions along which psychopathological phenomena could be classified to relate in a meaningful way to evoked potential activity. Such dimensions could be either more or less general than those customarily employed. There is little evidence, as of now, to indicate that more narrowly defined psychopathological categories would be more valid than those now used. On the other hand, there is some reason to suggest that more broadly based behavioral dimensions may be of value. Our own findings, with respect to psychological tests and psychiatric diagnosis (Chapters 8 and 9), suggest that our evoked-response measurements do correlate with severity of psychopathology. Not only do patients differ from nonpatients but, within a patient population, those with psychological test scores that indicate more severe psychopathology show evoked-potential findings that deviate more from the findings in nonpatients. The implication of these results is that broader concepts, such as severity of psychopathology, may yield dimensions that will be closer correlates of evoked potentials than more restricted ones, such as presence or absence of delusions. The numerous psychological correlates of evoked-potential measurements suggest that we must be tapping brain processes that are involved in many aspects of behavior. It may be that, to be appropriate, our conceptualized behavioral dimensions need to be broad in nature. One example of a broadly based behavioral dimension is consciousness, and this is perhaps the best established psychological correlate of brain electrical activity.

Evoked-potential research is becoming increasingly sophisticated as a consequence of both instrumental developments and increasing knowledge of parameters that must be controlled to avoid contamination of results. As these sophisticated techniques are applied, together with newer concepts that may emerge from attempts to view psychopathology from a different vantage point, the need for adequate theories of underlying neural mechanisms will become pressing. Such theories depend upon accumulation of a body of information that will bridge the gap between events taking place in the individual cell and the electrical signs, such as evoked potential, that can be recorded from the surface of the head. The major issues have been considered in a recent symposium of the Neurosciences Research Program (MacKay, 1969). The participants in this symposium surveyed the current status of the relationship between micro- and macroelectrode recordings. Although it was concluded that there is still a large gap between the two, it was felt that the gap promises to be both bridgeable and worth bridging.

One of the important possible ways of interpreting the surface-evoked response is illustrated by the work of Fox and O'Brien (1965). They demonstrated that a very high correlation can be obtained between the shape of the potential evoked by flash and the histogram relating the frequency of cell firing to the time after stimulus (poststimulus time histogram). The high correlation held for both positive and negative and early and late components of the flash evoked response. The data of Fox and O'Brien suggest that, at least under some stimulating conditions, the waveform of the average cortical evoked response may be interpreted as reflecting the distribution of firing rates in the neuronal population from which it was derived. Creutzfeldt *et al.* (1969) obtained evidence of a similar nature, which suggests a close correlation between the cortical surface response and the firing patterns of individual excitatory and inhibitory neurons in response to light flash. The participants in the Neurosciences Symposium, however, were agreed that evoked potentials could be understood in terms of simple experimental models like those suggested by the data of Fox and O'Brien and Creutzfeldt *et al.* only under limited circumstances.

Although hard knowledge is so far minimal, it is clear that a number of highly competent workers are engaged in attempts to unravel the problems of the basic neural mechanisms mediating evoked responses. The results obtained to date encourage some degree of hope that neural mechanisms underlying evoked-potential correlates of psychopathology may be forthcoming.

Mention should be made of one other area of investigation, which may be relevant to understanding psychopathology-related evoked-potential events. The studies in this area involve modification of evoked response characteristics by means of both operant and classical conditioning techniques in animals and man (Fox and Rudell, 1968; Rosenfeld *et al.*, 1969; Begleiter and Platz, 1969). The demonstrations that evoked-response components can be altered by conditioning procedures document the possibility that some aspects of response waveform may be determined as a consequence of experience.

Information bearing on the interpretation of evoked responses obviously is being sought by a variety of means and from several directions. It is unlikely that any one investigator can encompass all the different levels of evoked-potential research. The psychiatrist in this field can probably make his unique contribution at the level of helping to establish the best possible correlations between evoked-response characteristics and psychopathological events. If the correlations are good enough, his colleagues, working at other levels, will almost certainly provide fruitful explanatory hypotheses for them.

It seems possible to close this book on a note of cautious optimism. Although the past decade has taught us how difficult it is to apply evoked-potential methods in a valid and meaningful way to psychiatric problems, much positive information has been gained. The road ahead promises to be difficult, laborious, and lengthy, but fruitful. The yield of information garnered by psychiatrically oriented investigators of evoked responses should be significantly amplified by the findings of their colleagues in other branches of neurobiology.

References

Abrahamian, H.A., Allison, T., Goff, W.R., and Rosner, B.S. (1963). Effects of thiopental on human cerebral evoked responses. *J. Anesthesiol.* **24** : 650–657.

Akiyama, Y., Schulte, F.J., Schultz, M.A., and Parmelee, Jr., A.H. (1969). Acoustically evoked responses in premature and full term newborn infants. *Electroencephalog. clin. Neurophysiol.* **26** : 371–380.

Allison, T. (1962). Recovery functions of somatosensory evoked responses in man. *Electroencephalog. clin. Neurophysiol.* **114** : 331–343.

Allison, T., Goff, W.R., Abrahamian, H.A., and Rosner, B.S. (1962). Effects of barbiturate anesthesia upon human somatosensory responses. *Electroencephalog. clin. Neurophysiol.* **14** : 419.

Aserinsky, E., and Kleitman, N. (1965). Two types of ocular motility occurring in sleep. *J. Applied Physiol.* **8** : 1–10.

Barlow, J.S. (1957). An electronic method for detecting evoked responses of the brain and for reproducing their average waveforms. *Electroencephalog. clin. Neurophysiol.* **9** : 340–343.

Barlow, J.S. (1960). Rhythmic activity induced by photic stimulation in relation to intrinsic alpha activity of the brain in man. *Electroencephalog. clin. Neurophysiol.* **12** : 317–326.

Barlow, J.S. (1964). Evoked responses in relation to visual perception and oculomotor reaction times in man. *Ann. N. Y. Acad. Sci.* **112** : 432–467.

Barlow, J.S. (1964b). *Rhythmic Afterdischarges to Flashes.* Quarterly Progress Report 75, Research Laboratory of Electronics MIT (October, 1964), pp. 149–155.

Barnet, A.B., and Goodwin, R.S. (1965). Averaged evoked electroencephalographic responses to clicks in the human newborn, *Electroencephalog. clin. Neurophysiol.* **18** : 441–450.

Barnet, A.B., and Lodge, A. (1967). Click evoked EEG responses in normal and developmentally retarded infants. *Nature* **214** : 252–255.

Barron, F. (1953). An ego-strength scale which predicts response to psychotherapy. *J. Consult. Psychol.* **17** : 327–333.

Barry, W.M., and Ertl, J.P. (1966). Brain waves and human intelligence. *In:* F.G. Davis (ed.), *Modern Educational Developments: Another Look*, Educational Records Bureau, New York.

Bates, J.A.V. (1951). Electrical activity of the cortex accompanying movement. *J. Physiol.* **113** : 240–257.

Beatty, J., and Uttal, W.R. (1968). The effects of grouping visual stimuli on the cortical evoked potential. *Perception and Psychophysics* **4** : 214–216.

Beck, E., and Barolin, G.S. (1965). Effect of hypnotic suggestions on evoked potentials. *J. Nerv. Ment. Dis.* **140** : 154–161.

Beck, E., Dustman, R. E., and Beier, E. G. (1966). Hypnotic suggestions and visually evoked potentials. *Electroencephalog. clin. Neurophysiol.* **20** : 397–400.

Begleiter, H., and Platz, A. (1969). Evoked potentials: Modifications by classical conditioning. *Science* **166** : 760–771.

Bennett, P. B., Ackles, K. N., and Cripps, V. J. (1969). Effects of hyperbaric nitrogen and oxygen on auditory evoked responses in man. *Aerospace Med.* **40** : 521–525.

Bergamasco, B. (1966). Studio delle modificazioni della responsivita corticale nell'uomo indotte de farmaci ad azione sul SNC, *Sist. Nerv.* **18**:155–164.

Bergamasco, B. (1967). Modifications of cortical responsiveness in humans induced by drugs acting on the central nervous system. *Electroencephalog. clin. Neurophysiol.* **23** : 191.

Bergamasco, B., Bergamini, L., Mombelli, A. M., and Mutani, R. (1966). Longitudinal study of visual evoked potentials in subjects in posttraumatic coma. *Schweiz. Arch. Neurol. Neurochir. Psychiat.* **97** : 1–10.

Bergamini, L., and Bergamasco, B. (1967). *Cortical Evoked Potentials in Man,* Charles C. Thomas, Springfield.

Bergamini, L., Bergamasco, B., Mombelli, A. M., and Gandiglio, G. (1965). Visual evoked potentials in subjects with congenital aniridia. *Electroencephalog. clin. Neurophysiol.* **19** : 394–397.

Berger, H. Über das Elektrenkephalogramm des Menschen. *Arch. Psychiat.* **87** : 527–570.

Bickford, R. G., Jacobson, J. L., and Cody, D. T. (1964). Nature of average evoked potentials to sound and other stimuli in man. *Ann. N. Y. Acad. Sci.* **112** : 204–223.

Bigum, H. B., Dustman, R. E. and Beck, E. C. (1970). Visual and somatosensory evoked responses from mongoloid and normal children. *Electroencephalog. clin. Neurophysiol.* **28** : 576–585.

Bishop, M. P., Guthrie, M. B., and Hornsby, L. D. (1969). Objective and subjective tests in measuring response to CNS drugs in normal subjects. *J. Clin. Pharmacol.* **9** : 308–314.

Blacker, K. H., Jones, R. T., Stone, G. C., and Pfefferbaum, D. (1968). Chronic users of LSD: The "Acidheads." *Amer. J. Psychiat.* **125** : 341–351.

Bogacz, J., Vanzulli, A., and Garcia-Austt, E. (1962). Evoked responses in man. IV. Effects of habituation, distraction and conditioning upon auditory evoked responses. *Acta Neurol. Lat. Amer.* **8** : 244–252.

Bostem, F., Rousseau, J. C., Degossely, M., and Dongier, M. (1967). Psychopathological correlations of the non-specific portion of visual and auditory evoked potentials and the associated contingent negative variation. *Electroencephalog. clin. Neurophysiol.* Suppl. **26** : 131–138.

Brazier, M. A. B. (1958). Studies of evoked responses by flash in man and cat. *In:* H. H. Jasper, L. D. Proctor, R. S. Knighton, W. C. Noshay, and R. T. Costello (eds.), *Reticular Formation of the Brain,* Little, Brown and Co., Boston, pp. 151–168.

Brazier, M. A. B. (1960). Long persisting electrical traces in the brain of man and their possible relationship to higher nervous activity. *Electroencephalog. clin. Neurophysiol.* Suppl. **13** : 347–358.

Brazier, M. A. B. (1961). Introductory comments. *Electroencephalog. clin. Neurophysiol.* Suppl. **20** : 2–6.

Brazier, M. A. B. (1964). Evoked responses recorded from the depths of the human brain. *Ann. N. Y. Acad. Sci.* **112** : 33–59.

Brazier, M. A. B., and Finesinger, J. E., (1945). Action of barbiturates on the cerebral cortex. *Arch. Neurol. Psychiat.,* **53** : 51–58.

Brody, H. (1955). Organization of the cerebral cortex. III. A study of aging in the human cerebral cortex. *J. Comp. Neurol.* **102**:511–556.

Brown, B. B. (1968). Subjective and EEG responses to LSD in visualizer and non-visualizer subjects. *Electroencephalog. clin. Neurophysiol.* **25** : 372–379.

Brown, B. B. (1969). Effect of LSD on visually evoked responses to color in visualizer and non-visualizer subjects. *Electroencephalog. clin. Neurophysiol.* **27** : 356–363.

Brown, J. C. N., Shagass, C., and Schwartz, M. (1965). Cerebral evoked potential changes associated with the Ditran delirium and its reversal in man. *In:* J. Wortis (ed.), *Recent Advances in Biological Psychiatry,* Vol. VII, Plenum Press, New York, pp. 223–234.

Buchsbaum, M., and Silverman, J. (1968). Stimulus intensity control and the cortical evoked response. *Psychosom. Med.* **30** : 12–22.

Buchsbaum, M., and Silverman, J. (1970). Average evoked response and perception of the vertical. *J. Exp. Res. Personal.* **4** : 79–83.

Buller, A. J., and Styles, P. R. (1961). Improvement in signal/noise ratio with aid of a barrier-grid storage tube. *Proc. 3rd Int. Conf. Med. Electron.,* I. E. F., London, pp. 63–64.

Callaway, E. (1966). Averaged evoked responses in psychiatry. *J. Nerv. Ment. Dis.* **143** : 80–94.

Callaway, E. (1968). Invited discussion of Dr. C. Shagass' paper, *In:* D. H. Efron, J. O. Cole, J. Levine, and J. R. Wittenborn (eds.), *Psychopharmacology: A Review of Progress, 1952–1956,* U.S.P.S. Publication No. 1836, Washington, D. C., pp. 493–495.

Callaway, E. (1969). Diagnostic uses of the averaged evoked potentials. *In:* E. Donchin and D. B. Lindsley (eds.), *Average Evoked Potentials: Methods, Results and Evaluations,* National Aeronautics and Space Administration, Washington, D. C., pp. 299–311.

Callaway, E., and Buchsbaum, M. (1965). Effects of cardiac and respiratory cycles on averaged visual evoked responses. *Electroencephalog. clin. Neurophysiol.* **19** : 476–480.

Callaway, E., and Jones, R. M. (1970). Evoked responses for the study of complex cognitive functions *In:* M. Kretzman and J. Zubin (eds.), *Objective Indicators of Psychopathology,* in press.

Callaway, E., Jones, R. T., and Layne, R. S. (1965). Evoked responses and segmental set of schizophrenia. *Arch. Gen. Psychiat.* **12** : 83–89.

Calvet, J., and Scherrer, J. (1955). Des certaines limites et possibilités nouvelles en électro-physiologie. *In: Actes des Journées Mesure et Connaissance,* Rev. Métrologie, Paris, pp. 289–293

Cant, B. R., and Bickford, R. G. (1967). The effect of motivation on the contingent variation (CNV). *Electroencephalog. clin. Neurophysiol.* **23** : 594.

Caton, R. (1875). The electric currents of the brain. *Brit. Med. J.* **2** : 278

Cazzullo, C. S., Dubini, S., Lucioni, R., Monterisi, G. C., and Pietropolli Charmet (1967). Evaluation of photically evoked reponses in man by graphic superimposition, automatic averaging, and selected frequencies. *Electroencephalog. clin. Neurophysiol.* Suppl. **26** : 53–60.

Chalke, F. C. R., and Ertl, J. (1965). Evoked potentials and intelligence. *Life Sci.* **4** : 1319–1322.

Chapman, R. M. (1965). Evoked responses to relevant and irrelevant visual stimuli while problem solving. *Am. Psychol. Assn.* Proceedings, 73rd Annual Convention, pp. 177–178.

Chapman, R. M. (1969). *In:* E. Donchin and D. Lindsley (eds.), *Average Evoked Potentials. Methods, Results and Evaluations,* National Aeronautics and Space Administration, Washington, D. C., pp. 262–275.

Chapman, L. F., and Walter, R. D. (1965). Action of lysergic acid diethylamide on averaged human cortical evoked responses to light flash. *In:* J. Wortis (ed.), *Recent Advances in Biological Psychiatry,* Vol. VII, Plenum Press, New York, pp. 23–36.

Chapman, R. M., and Bragdon, H. R. (1964). Evoked responses to numerical visual stimuli while problem solving. *Nature* **203** : 1155–1157.

Ciganek, L. (1958). Postdécharge rythmique corticale chez l'homme évoquée par les stimuli photiques (Cortical rhythmic after-discharge in man evoked by photic stimulation). *Rev. Neurol.* **99** : 196–198.

Ciganek, L. (1959). The effect of largactil on the electroencephalographic response evoked potential to light stimulus in man. *Electroencephalog. clin. Neurophysiol.* **7** : 65–71.

Ciganek, L. (1961*a*). The EEG response (evoked potential) to light stimulus in man. *Electroencephalog. clin. Neurophysiol.* **13** : 165–172.

Ciganek L. (1961*b*). *Die electrenkephalographische Lichtreizantwort der menschlichen Hirnrinde,* Slowakische Akademie der Wissenschaften, Bratislava.

Ciganek, L. (1964). Excitability cycle of the visual cortex in man. *Ann. N. Y. Acad. Sci.* **112** : 241–253.

Cigánek, L. (1965). A comparative study of visual and auditory EEG responses in man. *Electro-encephalog. clin. Neuropsychiol.* **18** : 625–629.

Cigánek, L. (1967). The effects of attention and distraction on the visual evoked potential in man: A preliminary report. *Electroencephalog. clin. Neurophysiol.* Suppl. **26** : 70–73.

Claridge, G. S., and Herrington, R. N. (1960). Sedation threshold, personality, and the theory of neurosis. *J. Ment. Sci.* **106** : 1568–1583.

Clark, W. A. (1961). Digital techniques in neuroelectric data processing. *Electroencephalog. clin. Neurophysiol.* Suppl. **20** : 75–78.

Clark, W. A., Brown, R. M., Goldstein, M. H., Molnar, C. E., O'Brien, D. F., and Zieman, H. E. (1961). The average response computer (ARC): A digital device for computing averages and amplitude and time histograms of electrophysiological responses. *I.R.E. Trans. Bio-med. Electron.* **8** : 46–51.

Cleckley, H. (1959). Psychopathic states. *In:* S. Arieti (ed.), *American Handbook of Psychiatry,* Basic Books, New York, pp. 567–588.

Clynes, M. (1962). CAT: Computer of average transients. *Instrum. Control Syst.* **35** : 87–91.

Clynes, M. (1969). Dynamics of vertex evoked potentials: The R-M brain function. *In:* E. Donchin and D. Lindsley (eds.), *Average Evoked Potentials: Methods, Results, and Evaluations,* National Aeronautics and Space Administration, Washington, D.C., pp. 363–374.

Clynes, M., and Kohn, M. (1964). Specific responses of the brain to color stimuli. *Proc. 17th Ann. Conf. Engin. Med. Biol.* **32P.**

Clynes, M., and Kohn, M. (1968) Recognition of visual stimuli from the electric responses of the brain. *In:* N.S. Kline and E. Laska (eds.), *Computers and Electronic Devices in Psychiatry,* Grune and Stratton, New York, pp. 206–237.

Clynes, M., Kohn, M., and Lifshitz, K. (1964). Dynamics and spatial behavior of light evoked potentials, their modification under hypnosis, and on-line correlation in relation to rhythmic components. *Ann. N. Y. Acad. Sci.* **112** : 468–509.

Clynes, M., Kohn, M., and Gradijan, J. (1967). Computer recognition of the brain's visual perception through learning the brain's physiologic language. *I.E.E.E. International Convention Record,* Part **9** : 125–142.

Cobb, W. A., and Dawson, G. D. (1960). The latency and form in man of the occipital potentials evoked by bright flashes. *J. Physiol.* **152** : 108–121.

Cobb, W. A., and Morocutti, C. (eds.) (1967). The evoked potentials. *Electroencephalog. clin. Neurophysiol.* Suppl. 26.

Cohen, J. (1969). Very slow brain potentials relating to expectancy: the CNV. *In:* E. Donchin and D. B. Lindsley (eds.), *Average Evoked Potentials: Methods, Results, and Evaluations,* National Aeronautics and Space Administration, Washington, D.C., pp. 143–163.

Cohn, R. (1964). Rhythmic after activity in visual evoked responses. *Ann. N. Y. Acad. Sci.* **112** : 281–291.

Contamin, F., and Cathala, H. P. (1961). Réponses électro-corticales de l'homme normal éveille à des éclairs lumineux. Résultats obtenus à partir d'enregistrements sur le cuir chevelu, à l'aide d'un dispositif d'intégration. *Electroencephalog. clin. Neurophysiol.* **13** : 674–694.

Corletto, F., Gentilomo, A., Giannotti, M., Rosadini, G., Rossi, G. F., and Zattoni, J. (1966). A study of the relationship between degree of loss of consciousness, "spontaneous" EEG activity and "evoked" potentials following electro-shock. *Electroencephalog. clin. Neurophysiol.* **21** : 92.

Corssen, G., and Domino, E. F. (1964). Visually evoked responses in man: A method for measuring cerebral effects of preanesthetic medications. *Anesthesiology* **25** : 330–341.

Cracco, R. Q., and Bickford, R. G. (1968). Somatomotor and somatosensory evoked responses. Median nerve stimulation in man. *Arch. Neurol.* **18** : 52–68.

Creutzfeldt, O. D., and Kuhnt, U. (1967). The visual evoked potential: physiological, developmental and clinical aspects. *In:* W. Cobb and C. Morocutti (eds.), *The Evoked Potentials,* Elsevier Publishing Company, Amsterdam, pp. 29–41.

Creutzfeldt, O. D., Kugler, J., Morocutti, D., and Sommer-Smith, J. A. (1966a). Visual evoked potentials in normal human subjects and neurological patients. *Electroencephalog. clin. Neurophysiol.* **20** : 98–106.

Creutzfeldt, O.D., Rosina, A., Ito, M., and Probst, W. (1966b). Visual evoked response of single cells and of the EEG in primary visual area of the cat. *J. Neurophysiol.* **32** : 172–139.

Crighel, E., and Ciurea, E. (1966). Flash-evoked potentials in man. Variability of the evoked responses. *Electroencephalog. clin. Neurophysiol.* **21** : 99.

Dahlstrom, W.G., and Welsh, G.S. (1960). *MMPI Handbook: A Guide to Use in Clinical Practice and Research*, University of Minnesota Press, Minneapolis.

Dargent, J., and Dongier, M. (1969). *Variations Contingents Négatives*, University of Liège. Belgium.

Davis, H., Mast, T., Yoshie, N., and Zerlin, S. (1966). The slow response of the human cortex to auditory stimuli: recovery process. *Electroencephalog. clin. Neurophysiol.* **21** : 105–113.

Dawson, G.D. (1947). Cerebral responses to electrical stimulation of peripheral nerve in man. *J. Neurol. Neurosurg. Psychiat.* **10** : 134–140.

Dawson, G.D. (1951). A summation technique for detecting small signals in a large irregular background. *J. Physiol.* **115** : 2P–3P.

Dawson, G.D. (1954). A summation technique for the detection of small evoked potentials. *Electroencephalog. clin. Neurophysiol.* **6** : 65–84.

Dawson, G.D. (1956). The relative excitability and conduction velocity of sensory and motor nerve fibres in man. *J. Physiol.* **131** : 436–451.

Debecker, J., Desmedt, J.E., and Manil, J. (1965). On the relationship between the threshold of tactile perception and the somatosensory cortical evoked potentials in man. *C.R. Acad. Sci.* **260** : 687–689.

DeLange, H. (1958). Research into the dynamic nature of the human foveacortex systems with intermittent and modulated light. I. Attenuation characteristics with white and colored light. *J. Opt. Soc. Amer.* **48** : 777–784.

Dement, W.C. (1967). Sleep and dreams. *In:* A.M. Freedman and H.I. Kaplan (eds.), *Comprehensive Textbook of Psychiatry*, Williams and Wilkins Company, Baltimore, pp. 77–88.

Dement, W.C., and Kleitman, N. (1957). Cyclic variations in EEG during sleep and their relation to eye movements, body motility, and dreaming. *Electroencephalog. clin. Neurophysiol.* **9** : 673–690.

Desmedt, J.E., Manil, J., Chorazyna, H., and Debecker, J. (1967). Potentiel évoqué cérébral et conduction corticipète pour une volée d'influx somathésique chez le nouveau-né normal. *C.R. Soc. Biol.* **161** : 205–209.

Diamond, S.P. (1964). Input–output relations. *Ann. N.Y. Acad. Sci.* **112** : 160–171.

Dolce, G., and Kaemmerer, E. (1967). Effect of the Benzodiazepin adumbran on the resting and sleep EEG, and on the visual evoked potential in adult man. *Med. Welt* **67** : 510–514.

Domino, E.F. (1967). Effects of preanesthetic and anesthetic drugs on visually evoked responses. *Anesthesiology* **28** : 184–191.

Domino, E.F., and Corssen, G. (1964). Visually evoked responses in anesthetized man with and without induced muscle paralysis. *Ann. N.Y. Acad. Sci.* **112** : 226–237.

Domino, E.F., Corssen, G., and Sweet, R.G. (1963). Effects of various general anesthetics on the visually evoked responses in man. *Anesth. Analog.* **42** : 735–747.

Domino, E.F., Matsuoka, S., Waltz, J., and Cooper, I.S. (1964). Simultaneous recordings of scalp and epidural somatosensory evoked responses in man. *Science* **145** : 1199–1200.

Domino, E.F., Matsuoka, S., Waltz, J., and Cooper, I.S. (1965). Effects of cryogenic thalamic lesions on the somesthetic evoked response in man. *Electroencephalog. clin. Neurophysiol.* **19** : 127–138.

Donchin, E. (1966). A multivariate approach to the analysis of averaged evoked potentials. *I.E.E.E. Trans. Biomed. Eng.* **13** : 131–139.

Donchin, E. (1969). Data analysis techniques in average evoked potential research. *In:* E. Donchin and D. Lindsley, (eds.), *Average Evoked Potentials: Methods, Results, and Evaluations*, National Aeronautics and Space Administration, Washington, D.C., pp. 199–236.

Donchin, E., and Lindsley, D. B. (1966). Averaged evoked potentials and reaction times to visual stimuli. *Electroencephalog. clin. Neurophysiol.* **20** : 217–223.

Donchin, E., and Lindsley, D. B. (1969). *Average Evoked Potentials: Methods, Results and Evaluations,* National Aeronautics and Space Administration, Washington, D. C.

Dongier, M. (1969). *In:* J. Dargent and M. Dongier (eds.), *Variations Contingentes Négatives,* University of Liège, pp. 90–91.

Dongier, M., and Bostem, M. F. (1967). Essais d'application en psychiatrie de la Variation Contingente Négative. *Acta Neurol. Belg.* **67** : 640–645.

Dustman, R. E., and Beck, E. C. (1963). Long-term stability of visually evoked potentials in man. *Science* **142** : 1480–1481.

Dustman, R. E., and Beck, E. C. (1965a). The visually evoked potential in twins. *Electroencephalog. clin. Neurophysiol.* **19** : 570–575.

Dustman, R. E., and Beck, E. C. (1965b). Phase of alpha brain waves, reaction time and visually evoked potentials. *Electroencephalog. clin. Neurophysiol.* **18** : 433–440.

Dustman, R. E., and Beck, E. C. (1966). Visually evoked potentials: amplitude changes with age. *Science* **151** : 1013–1015.

Dustman, R. E., and Beck, E. C. (1969). The effects of maturation and aging on the waveform of visually evoked potentials. *Electroencephalog. clin. Neurophysiol.* **26** : 2–11.

Eason, R. G., and White, C. T. (1967). Averaged occipital responses to stimulation of sites in the nasal and temporal halves of the retina. *Psychon. Sci.* **7** : 309–310.

Eason, R. G., Aiken, L. R., White, C. T., and Lichtenstein, M. (1964). Activation and behavior. II. Visually evoked cortical potentials in man as indicants of activation level. *Percept. Motor Skills* **19** : 875–895.

Eason, R. G., Groves, P., White, C. T., and Oden, D. (1967a). Evoked cortical potentials: Relation to visual field and handedness. *Science* **156** : 1643–1646.

Eason, R. G., Oden, D., and White, C. T. (1967b). Visually evoked cortical potentials and reaction time in relation to site of retinal stimulation. *Electroencephalog. clin. Neurophysiol.* **22** : 313–324.

Eason, R. G., Harter, M. R., and White, C. T. (1969). Effects of attention and arousal on visually evoked cortical potentials and reaction time in man. *Physiol. Behav.* **4** : 283–289.

Ebe, M. Mikami, T., Aki, M., and Miyazaki, M. (1962). Electrical responses evoked by photic stimulation in human cerebral cortex. *Tohoku J. Exper. Med.* **77** : 353–366.

Ebe, M., Meier-Ewert, K. H., and Broughton, R. (1969). Effects of intravenous diazepam (Valium) upon evoked potentials of photosensitive epileptic and normal subjects. *Electroencephalog. clin. Neurophysiol.* **27** : 429–435.

Efron, R. (1964). Artificial synthesis of evoked responses to light flash. *Ann. N. Y. Acad. Sci.* **112** : 292–304.

Ellingson, R. J. (1960). Cortical electrical responses to visual stimulation in the human infant. *Electroencephalog. clin. Neurophysiol.* **12** : 663–677.

Ellingson, R. J. (1964). Cerebral electrical responses to auditory and visual stimuli in the infant (Human and subhuman studies). *In:* P. Kellaway and I. Petersén (eds.), *Neurological and Electroencephalographic Correlative Studies in Infancy,* Grune and Stratton, New York, pp. 78–116.

Ellingson, R. J. (1966). Relationship between EEG and intelligence: a commentary. *Psychol. Bull.* **65** : 91–98.

Emde, J. (1964). A time locked low level calibrator. *Electroencephalog. clin. Neurophysiol.* **16** : 616–618.

Emde, J. W., and Shipton, H. W. (1970). A digitally controlled constant current stimulator. *Electroencephalog. clin. Neurophysiol.* **29** : 310–313.

Ennever, J., Gartside, I. B., Lippold, O. C. J., Novotny, G. E. K., and Shagass, C. (1967). Contamination of the human cortical evoked response with potentials of intra-orbital origin. *J. Physiol.* **191** : 6–7.

Ertl, J., and Schafer, E. W. P. (1967). Cortical activity preceding speech. *Life Sci.* **6** : 437–479.

Ertl, J., and Schafer, E. W. P. (1969). Brain response correlates of psychometric intelligence. *Nature* 223 : 421–422.

Ervin, F. R., and Mark, V. H. (1964). Studies of the human thalamus: IV. Evoked responses. *Ann. N. Y. Acad. Sci.* 112 : 81–92.

Evans, C. C. (1953). Spontaneous excitation of the visual cortex and association areas—lambda waves. *Electroencephalog. clin. Neurophysiol.* 5 : 69–74.

Eysenck, H. J. (1959). *Manual of the Maudsley Personality Inventory*, University of London Press Ltd., London.

Favale, E., Loeb, C., and Manfredi, M. (1964). Responses evoked by stimulation of the visual pathways during natural sleep and during arousal. *Electroencephalog. clin. Neurophysiol.* 17 : 584.

Ferriss, G. S., Davis, G. D., Dorsen, M. McF., and Hackett, E. R.(1967a). Maturation of the evoked response to auditory stimuli in human infants. *Electroencephalog. clin. Neurophysiol.* 23 : 83.

Ferriss, G. S., Davis, G. D., Dorsen, M. McF., and Hackett, E. R. (1967b). Changes in latency and form of the photically induced averaged evoked response in human infants. *Electroencephalog. clin. Neurophysiol.* 22 : 305–312.

Fischel, H. (1969). Visual evoked potentials in prematures, newborns, infants and children by stimulation with colored light. *Electroencephalog. clin. Neurophysiol.* 27 : 660.

Floris, V., Morocutti, C., Amabile, G., Bernardi, G., Rizzo, P. A., and Vasconetto, C. (1967). Recovery cycle of visual evoked potentials in normal and schizophrenic subjects. *Electroencephalog. clin. Neurophysiol.* Suppl. 26 : 74–81.

Floris, V., Morocutti, C., Amabile, G., Bernardi, G., and Rizzo, P. A. (1968). Recovery cycle of visual evoked potentials in normal, schizophrenic and neurotic patients. *In:* N. S. Kline and E. Laska (eds.), *Computers and Electronic Devices in Psychiatry*, Grune and Stratton, New York, pp. 194–205.

Floris, V., Morocutti, C., Amabile, G., Bernardi, G., and Rizzo, P. A. (1969). Cerebral reactivity in psychiatric and epileptic patients. *Electroencephalog. clin. Neurophysiol.* 27 : 680.

Flugel, F., and Itil, T. (1962). Klinisch-elektrenkephalographische Untersuchungen mit Verwirrtheit hervorrufenden Substanzen. *Psychopharmacologia* 3 : 79–98.

Fox, S. S., and O'Brien, J. H. (1965). Duplication of evoked potential waveform by curve of probability of firing of a single cell. *Science* 147 : 888–890.

Fox, S. S., and Rudell, A. (1968). Operant controlled neural event: formal and systematic approach to electrical coding of behavior in brain. *Science*, 162 : 1299.

Garcia-Austt, E. (1963). Influence of the state of awareness upon sensory evoked potentials. *Electroencephalog. clin. Neurophysiol.* Suppl. 24 : 76–89.

Garcia-Austt, E., Vanzulli, A., Bogacz, J., and Rodriguez-Barrios, R. (1963). Influence of the ocular muscles upon photic habituation in man. *Electroencephalog. clin. Neurophysiol.* 15 : 281–286.

Garcia-Austt, E., Bogacz, J., and Vanzulli, A. (1964). Effects of attention and inattention upon visual evoked response. *Electroencephalog. clin. Neurophysiol.* 17 : 136–143.

Gartside, I. B., Lippold, O. C. J., and Meldrum, B. S. (1966). The evoked cortical somatosensory response in normal man and its modification by oral lithium carbonate. *Electroencephalog. clin. Neurophysiol.* 20 : 382–390.

Gasser, H. S., and Grundfest, H. (1936). Action and excitability in mammalian A fibers. *Amer. J. Physiol.* 117 : 113–133.

Gastaut, H., and Régis, H. (1965). The visually evoked potentials recorded transcranially in man. *In:* L. D. Proctor and W. R. Adey (eds.), *NASA Symposium on the Analysis of Central Nervous System and Cardiovascular Data Using Computer Methods*, NASA, Sp-72, Washington, D. C., pp. 7–34.

Gastaut, H., Gastaut, Y., Roger, A., Corriol, J., and Naquet, R. (1951a). Étude électrographique du cycle d'excitabilité cortical. *Electroencephalog clin. Neurophysiol.* 3 : 401–428.

Gastaut, H., Carriol, J., and Roger, A. (1951b). Le cycle d'excitabilité des systèmes afferent corticaux chez l'homme. *Rev. Neurol.* 84 : 602–605.

Gastaut, H., Beek, E., Faidherbe, J., Franck, G., Fressy, J., Rémond, A., Smith, C., and

Werre, P. (1963). A transcranial chronographic and topographic study of cerebral potentials by photic stimulation in man. *In:* G. Moruzzi, A. Fessard, and H. H. Jasper (eds.), *Progress in Brain Research, Vol. 1, Brain Mechanisms,* Elsevier Publishing Company, Amsterdam, pp. 374–392.

Gastaut, H., Orfanos, A., Poiré, R., Régis, H., Saier, J., and Tassinari, C.A. (1966). Effets de l'adaptation à l'obscurité sur les potentiels évoqués visuels de l'homme (Effects of dark adaptation on visual evoked potentials in man). *Rev. Neurol.* **36** : 63–72.

Gastaut, H., Bostem, F., Waltregny, A., Poiré, R., and Régis, H. (1967). *"Les Activités Electriques Cérébrales Spontanées et Evoquées chez l'Homme,"* Gauthier-Villars, Paris.

Gershon, S., and Olariu, J. (1960). JB 329—A new psychotomimetic: Its antagonism by tetrahydroaminacrin and its comparison with LSD, mescaline and sernyl. *J. Neuropsychiat.* **1** : 283–292.

Gibbs, F.A., Davis, H., and Lennox, W.G. (1935). The electroencephalogram in epilepsy and in conditions of impaired consciousness. *Arch. Neurol. Psychiat.* **34** : 1133–1148.

Giblin, D.R. (1964). Somatosensory evoked potentials in healthy subjects and in patients with lesions of the nervous system. *Ann. N. Y. Acad. Sci.* **112** : 93–142.

Gilden, L., Vaughan, Jr., H.G., and Costa, L.D. (1966). Summated human EEG potentials with voluntary movement. *Electroencephalog. clin. Neurophysiol.* **20** : 433–438.

Goff, W.R., Rosner, B.S., and Allison, T. (1962). Distribution of cerebral somatosensory evoked responses in normal man. *Electroencephalog. clin. Neurophysiol.* **14** : 697–713.

Goff, W.R., Allison, T., Shapiro, A., and Rosner, B.S. (1966). Cerebral somatosensory responses evoked during sleep in man. *Electroencephalog. clin. Neurophysiol.* **21** : 1–9.

Goff, W.R., Matsumiya, T., Allison, T., and Goff, G.D. (1969). Cross-modality comparisons of averaged evoked potentials. *In:* E. Donchin and D. Lindsley (eds.), *Average Evoked Potentials: Methods, Results, and Evaluations,* National Aeronautics and Space Administration, Washington, D.C., pp. 95–141.

Goldfarb, A.I. (1967). Geriatric psychiatry. *In:* A.M. Freedman and H.I. Kaplan (eds.), *Comprehensive Textbook of Psychiatry,* Williams and Wilkins, Baltimore, pp. 1564–1587.

Goldstein, L., and Beck, R.A. (1965). Amplitude analysis of the electroencephalogram. *Internat. Rev. Neurobiol.* **8** : 265–312.

Gottschaldt, K. (1929). Über den Einfluss der Erfahrung auf die Wahrnehmung von Figuren, II. *Psychol. Forschung* **12** : 1–88.

Gross, M.M., Tobin, M., Kissin, B., Halpert, E., and Sabot, L. (1964). Evoked responses to clicks in delirium tremens: A preliminary report. *Ann. N.Y. Acad. Sci.* **112** : 543–546.

Gross, M.M., Begleiter, H, Tobin, M., and Kissin, B. (1965). Auditory evoked response comparison during counting clicks and reading. *Electroencephalog. clin. Neurophysiol.* **18** : 451–454.

Gross, M.M., Begleiter, H., Tobin, M., and Kissin, B. (1966). Changes in auditory evoked response induced by alcohol. *J. Nerv. Ment. Dis.* **143** : 152–156.

Grundfest, H. (1959). Synaptic and ephaptic transmission. *In:* J. Field, H.W. Magoun, and V.E. Hall (eds.), *Handbook of Physiology, Section* 1, *Neurophysiology,* Vol. 1, American Physiological Society, Washington, D.C., pp. 147–197.

Guerrero-Figueroa, R., and Heath, R.G. (1964). Evoked responses and changes during attentive factors in man. *Arch. Neurol.* **10** : 74–84.

Haider, M. (1967). Vigilance, attention, expectation and cortical evoked potentials. *Acta Psychologica* **27** : 245–252.

Haider, M., Spong, P., and Lindsley, D.B. (1964). Attention, vigilance and cortical evoked potentials in humans. *Science* **145** : 180–182.

Hakerem, G., and Lidsky, A. (1970). Characteristics of pupillary reactions in psychiatric patients and normals. *In:* M. Kietzman and J. Zubin (eds.), *Objective Indicators of Psychopathology,* in press.

Halliday, A.M., and Mason, A.A. (1964). The effect of hypnotic anesthesia on cortical responses. *J. Neurol. Neurosurg. Psychiat.* **27** : 300–312.

Halliday, A.M., and Wakefield, G.S. (1963). Cerebral evoked potentials in patients with dissociated sensory loss. *J. Neurol. Neurosurg. Psychiat.* **26** : 211–219.

Harris, J.A., and Bickford, R.G. (1967). Cross-sectional plotting of EEG potential fields. *Electroencephalog. clin. Neurophysiol.* **23** : 88–89.

Häseth, K., Shagass, C., and Straumanis, J.J. (1969). Perceptual and personality correlates of EEG and evoked response measures. *Biol. Psychiat.* **1** : 49–60.

Helmchen, H., and Kunkel, M. (1964). Befunde zur rhythmischen Nachschwankung bei optisch ausgelösten Reizantworten (evoked responses) im EEG des Menschen. *Arch. Psychiat. und Z. ges. Neurol.* **205** : 397–408.

Heninger, G. R. (1969a). Lithium effects on cerebral cortical function in manic depressive patients. *Electroencephalog. clin. Neurophysiol.* **27** : 670.

Heninger, G.R. (1969b). Central neurophysiologic correlates of depressive symptomatology. Presented at NIMH Workshop on the Psychobiology of the Depressive Illnesses, in press.

Heninger, G. R., and Speck, L. (1966). Visual evoked responses and mental status of schizophrenics. *Arch. Gen. Psychiat.* **15** : 419–426.

Heninger, G.R., McDonald, R.K., Goff, W.R., and Sollberger, A. (1969). Diurnal variations in the cerebral evoked response and EEG: Relations to 17-hydroxycorticosteroid levels. *Arch. Neurol.* **21** : 330–337.

Hernández-Peón, R., and Donoso, M. (1959). Influence of attention and suggestion upon subcortical evoked electrical activity in the human brain. *In:* L. van Bogaert and J. Radermecker (eds.), *First International Congress on Neurological Sciences,* Vol. 3, Pergamon Press, New York, pp. 385–396.

Hillyard, S.A. (1969a). The CNV and the vertex evoked potential during signal detection: A preliminary report. *In:* E. Donchin and D. Lindsley (eds.), *Average Evoked Potentials: Methods, Results, and Evaluations,* National Aeronautics and Space Administration, Washington, D.C., pp. 349 353.

Hillyard, S.A. (1969b). *In:* J. Dargent and M. Dongier (eds.), *Variations Contingentes Négatives,* University of Liège, pp. 79–81, 103–111.

Hillyard, S.A. (1969c). Relationships between the contingent negative variation (CNV) and reaction time. *Physiol. Behav.* **4** : 351–358.

Hinsie, L.E., and Campbell, R.J. (1960). *Psychiatric Dictionary,* Oxford University Press, New York.

Hirsch, J.F., Pertuiset, B., Calvet, J., Buisson-Ferey, J., Fischgold, H., and Scherrer, J. (1961). Étude des réponses électrocorticales obtenues chez l'homme par des stimulations somesthésiques et visuelles. *Electroencephalog. clin. Neurophysiol.* **13** : 411–424.

Hrbek, A., Hrbkova, M., and Lenard, H. (1968). Somatosensory evoked responses in newborn infants. *Electroencephalog. clin. Neurophysiol.* **25** : 443–448.

Ikuta, T. (1966). Effects of Thiopental on the human somatosensory evoked response. *Folia Psychiat. Neurol. Jap.* **20** : 18–31.

Irwin, D.A., Knott, J.R., McAdam, D.W., and Rebert, C.S. (1966). Motivational determinants of the "contingent negative variation," *Electroencephalog. clin. Neurophysiol.* **21** : 538–543.

Ivanitsky, A.M., and Arzeulova, O.K. (1968). The use of evoked potentials for the analysis of the interrelation of the ascending projection systems in reactive psychosis. *Activitas Nervosa Superior 10, 2* : 222.

Jasper, H.H. (1958). The ten twenty electrode system of the International Federation. *Electroencephalog. clin. Neurophysiol.* **10** : 371–375.

Jasper, H., Lende, R., Rasmussen, T. (1960). Evoked potentials from the exposed somatosensory cortex in man. *J. Nerv. Ment. Dis.* **130** : 526–537.

Jewett, D.L., Romano, M.N., and Williston, J.S. (1970). Human auditory evoked potentials: possible brain stem components detected on the scalp. *Science* **167**: 1517–1518.

John, E.R., Ruchkin, D.S., and Villegas, J. (1964). Experimental background: Signal analysis and behavioral correlates of evoked potential configurations in cats. *Ann. N.Y. Acad. Sci.* **112** : 362–420.

Jones, J., Itil, T.M., Keskiner, A., Holden, J.M.C., and Ulett, G.A. (1969). Psychotropic drug-induced alterations of somatosensory evoked responses in schizophrenic patients. *Excerpta Med. Int. Cong. Ser.* **180** : 338–391.

Jones, R. T., and Callaway, E. (1970). Auditory evoked responses in schizophrenia. *Biol. Psychiat.* **2** : 291–298.

Jones, R. T., Blacker, K. H., Callaway, E., and Layne, R. S. (1965). The auditory evoked response as a diagnostic and prognostic measure in schizophrenia. *Amer. J. Psychiat.* **122** : 33–41.

Jones, R. T., Blacker, K. H., and Callaway, E. (1966). Perceptual dysfunction in schizophrenia: Clinical and auditory evoked response findings. *Amer. J. Psychiat.* **123** : 639–645.

Kahneman, D., and Peavler, W. S. (1969). Incentive effects and pupillary changes in association learning. *J. Exp. Psychol.* **79** : 312–318.

Kamiya, J. (1968). Conscious control of brain waves. *Psychology Today* **1** : 57–60.

Kassabgui, M., Cadilhac, J., and Passouant, P. (1966). Studies on evoked responses in waking and during sleep in young children. *Rev. Neurol.* **115** : 84–88.

Katzman, R. (1964a). *Sensory Evoked Responses in Man*, Vol. 112, Annals of New York Academy of Science, New York, pp. 3–4.

Katzman, R. (1964b). The validity of the visual evoked response in man. *Ann. N. Y. Acad. Sci.* **112** : 238–240.

Knott, J. R., and Irwin, D. A. (1968). Anxiety, stress and the contingent (CNV) negative variation, *Electroencephalog. clin. Neurophysiol.* **24** : 286–287.

Köhler, W., Held, R., and O'Connell, D. N. (1952). An investigation of cortical currents. *Proc. Amer. Phil. Soc.* **96** : 290–330.

Kooi, K. A., and Bagchi, B. K. (1964a). Observations on early components of the visual evoked response and occipital rhythms. *Electroencephalog. clin. Neurophysiol.* **17** : 638–643.

Kooi, K. A., and Bagchi, B. K. (1964b). Visual evoked responses in man: normative data. *Ann. N. Y. Acad. Sci.* **112** : 254–269.

Kooi, K. A., Bagchi, B. K., and Jordan, R. N. (1964). Observations on photically evoked occipital and vertex waves during sleep in man. *Ann. N. Y. Acad. Sci.* **112** : 270–280.

Kopec, J. (1967). Transistor suppressor of electric network interferences and its application during electroencephalographic registration. *Polish Med. J.* **6** : 996–1002.

Kopell, B. S., Wittner, W. K., Lunde, D., Warrick, G., and Edwards, D. (1970a). Cortical effects on averaged evoked potential, alpha-rhythm, time estimation, and two-flash fusion threshold. *Psychosom. Med.* **32** : 39–49.

Kopell, B. S., Wittner, W. K., Lunde, D., Warrick, G., and Edwards, D. (1970b). Influence of Triiodothyronine on selective attention in man as measured by the visual averaged evoked potential. *Psychosom. Med.* **32** : 495–502.

Kornhüber, H. H., and Deecke, L. (1965). Cerebral potential changes in voluntary and passive movements in man: readiness potential and reafferent potential. *Pflügers Arch. ges. Physiol.* **284**: 1–17.

Kozhevnikov, V. A. (1958). Photo-electric method of selecting weak electrical responses of the brain. *Sechenov J. Physiol.* **44** : 801–809.

Krönlein, R. U. (1898). Zur cranio-cerebralen Topographie. *Beitr. klinisch. Chirurg.* **22** : 364–370.

Lader, M. H., and Norris, II. (1968). Effect of nitrous oxide on the auditory evoked response in man. *Nature* **218** : 1081–1082.

Laget, P., Raimbault, J., Lortholary, O., and Hanin-Ferrey, D. (1967). Evolution des potentiels évoqués somesthésiques chez le nourrisson et l'enfant. Premiers résultats. *C. R. Soc. Biol.* **161** : 1235–1242.

Lairy, G. C., and Guibal, M. (1969). Données préliminaires concernant l'étude de l'onde d'expectative chez l'enfant déficient visuel. *In:* J. Dargent and M. Dongier (eds.), *Variations Contingentes Négatives*, University of Liège, pp. 177–190.

Landis, C. (1954). Determinants of the critical flicker-fusion threshold, *Physiol. Rev.* **34** : 259–286.

Levonian, E. (1966). Evoked potential in relation to subsequent alpha frequency. *Science* **152** : 1280–1282.

Lewis, E.G., Dustman, R.E., and Beck, E.C. (1970). The effects of alcohol on visual and somatosensory evoked responses. *Electroencephalog. clin. Neurophysiol.* **28** : 202–205.

Liberson, W.T. (1966). Study of evoked potentials in aphasics. *Amer. J. Physical Med.* **45** : 135–142.

Liberson, W.T., and Kim, K.C. (1963). Mapping evoked potentials elicited by stimulation of median and peroneal nerves. *Electroencephalog. clin. Neurophysiol.* **15** : 721.

Libet, B., Alberts, W.W., Wright, E.W., Jr., and Feinstein, B. (1967). Responses of human somatosensory cortex to stimuli for conscious sensation. *Science* **158** : 1597–1600.

Lifshitz, K. (1966). The averaged evoked cortical response to complex visual stimuli. *Psychophysiology* **3** : 55–68.

Lifshitz, K. (1969). An examination of evoked potentials as indicators of information processing in normal and schizophrenic subjects. *In:* E. Donchin and D.B. Lindsley (eds.), *Average Evoked Potentials: Methods, Results, and Evaluations,* National Aeronautics and Space Administration, Washington, D.C., pp. 318–319, 357–362.

Lille, F., Lerique, A., Pottier, M., Scherrer, J., and Thieffry, S. (1968). Cortical evoked responses during coma in children. *Presse Med.* **76** : 1411–1414.

Loomis, A.L., Harvey, E.N., and Hobart, G.A. (1935). Potential rhythms of the cerebral cortex during sleep. *Science* **81** : 597–598.

Loomis, A.L., Harvey, E.N., and Hobart, G.A. (1938). Distribution of disturbance patterns in the human electroencephalogram, with special reference to sleep. *J. Neurophysiol.* **1** : 413.

Low, M.D. (1969). *In:* J. Dargent and M. Dongier (eds.), *Variations Contingentes Négatives,* University of Liége, Belgium, pp. 81–83.

Low, M.D., and McSherry, J.W. (1968). Further observations of psychological factors involved in CNV genesis. *Electroencephalog. clin. Neurophysiol.* **25** : 203–207.

Low, M.D., Frost, J.D., Borda, R.P., and Kellaway, P. (1966). Surface negative slow potential shift associated with conditioning in man. *Neurology* **16** : 771–782.

Low, M.D., Coats, A.C., Rettig, G.M., and McSherry, J.W. (1967). Anxiety, attentiveness-alertness: A phenomenological study of the CNV. *Neuropsychologica* **5** : 379–384.

MacKay, D.M. (1969). Evoked Brain Potentials as Indicators of Sensory Information Processing. *Neurosciences Res. Prog. Bull.* **7**, No. 3.

Malerstein, A.J., and Callaway, E. (1969). Two-tone average evoked response in Korsakoff patients. *J. Psychiat. Res.* **6** : 253–260.

Manil, J., Desmedt, J.E., Debecker, J., and Chorazyna, H. (1967). Les potentiels cérébraux évoqués par la stimulation de la main chez le nouveau-né normal. *Revue Neurologique* **117**: 53–61.

Mayer-Gross, W., Slater, E., and Roth, M. (1954). *Clinical Psychiatry,* Cassell, London.

McAdam, D.W. (1966). Slow potential changes recorded from human brain during learning of a temporal interval. *Psychon. Sci.* **6** : 435–436.

Mc-Adam, D.W. (1969). Increases in CNS excitability during negative cortical slow potentials in man. *Electroencephalog. clin. Neurophysiol.* **26** : 216–219.

McAdam, D.W., and Seales, D.M. (1969). Bereitschaftspotential enhancement with increased level of motivation. *Electroencephalog. clin. Neurophysiol.* **27**: 73–75.

McAdam, D.W., Irwin, D.A., Rebert, C.S., and Knott, J.R. (1966). Cognative control of the contingent negative variation. *Electroencephalog. clin. Neurophysiol.* **21** : 194–195.

McCallum, W.C., and Walter, W.G. (1968). The effects of attention and distraction on the contingent negative variation in normal and neurotic subjects. *Electroencephalog. clin. Neurophysiol.* **25**: 319–329.

Meduna, L.J., and Abood, L.J. (1959). Studies of a new drug (Ditran) in depressive states. *J. Neuropsychiat.* **1** : 20–22.

Meehl, P.E. and Dahlstrom, W.G. (1960). Objective configural rules for discriminating psychotic from neurotic MMPI profiles. *J. Consult. Psychol.* **24** : 375–387.

Monnier, M., and Rozier, J. (1968). Retinal and cortical evoked responses (ON and OFF) to isoenergetic colour stimuli in man. *In:The Clinical Value of Electroretinography,* Karger, Basel/New York, pp. 95–109.

Morrell, L. K., and Morrell, F. (1966). Evoked potentials and reaction times: A study of intra-individual variability. *Electroencephalog. clin. Neurophysiol.* **20** : 567–575.

Näätänen, R. (1967). Selective attention and evoked potentials. *Ann. Acad. Sci. Fennicae Helsinki,* **151** : 1–226.

Nākagawa, Y. (1965). Studies on evoked potentials to click and somatosensory stimulation in the waking state and during sleep in man. *Folia Psychiat. Neurol. Jap.* **19** : 279–293.

Nishitani, H., and Kooi, K. A. (1968). Cerebral evoked responses in hypothyroidism. *Electroencephalog. clin. Neurophysiol.* **24** : 554–560.

Norris, A. H., Shock, N. W., and Wagman, I. H. (1953). Age changes in the maximum conduction velocity of motor fibers of human ulnar nerves, *J. Appl. Physiol.* **5** : 589–593.

Ojemann, G. A., and Henkin, R. I. (1967). Steroid dependent changes in human visual evoked potentials. *Life Sci.* **6** : 327–334.

Ormsby, J. F. A. (1961). Design of numerical filters with application to missile data processing. *J. Ass. Comp. Mach.* **8** : 440–446.

Ornitz, E. M. (1969). A note on the AEP of autistic children recorded during sleep. *In:* E. Donchin and D. B. Lindsley (eds.), *Average Evoked Potentials: Methods, Results, and Evaluations,* National Aeronautics and Space Administration, Washington, D. C., pp. 355–356.

Ornitz, E. M., and Ritvo, E. R. (1968). Neurophysiologic mechanisms underlying perceptual inconstancy in autistic and schizophrenic children. *Arch. Gen. Psychiat.* **19** : 22–27.

Ornitz, E. M., Ritvo, E. R., Carr, E. M., Panman, L. M., and Walter, R. D. (1967). The variability of the auditory averaged evoked response during sleep and dreaming in children and adults. *Electroencephalog. clin. Neurophysiol.* **22** : 514–524.

Ornitz, E. M., Ritvo, E. R., Panman, L. M., Lee, Y. H., Carr, E. M., and Walter, R. D. (1968). The auditory evoked response in normal and autistic children during sleep. *Electroencephalog. clin. Neurophysiol.* **25** : 221–230.

Overton, D. A., and Shagass, C. (1969). Distribution of eye movement and eyeblink potentials over the scalp. *Electroencephalog. clin. Neurophysiol.* **27** : 546.

Pascal, G. R., and Suttell, B. J. (1951). *The Bender–Gestalt Test: Quantification and Validity for Adults,* Grune and Stratton, New York.

Perez-Borja, C., Chatrian, G. E., Tyce, F. A., and Rivers, M. H. (1962). Electrographic patterns of the occipital lobe in man: A topographic study based on use of implanted electrodes. *Electroencephalog. clin. Neurophysiol.* **14** : 171–182.

Perry, N. W., and Childers, D. G. (1969). *The Human Visual Evoked Response. Method and Theory,* Charles C Thomas, Springfield.

Perry, N. W., Jr., and Copenhaver, R. M. (1965). Differential cortical habituation with stimulation of central and peripheral retina. *Percept. Motor Skills* **20** : 1209–1213.

Petrie, A. (1967). *Individuality in Pain and Suffering,* University of Chicago Press, Chicago.

Poiré, R., Tassinari, C. A., Régis, H., and Gastaut, H. (1967). Effects of diazepam (Valium) on the responses evoked by light stimuli in man (lambda waves, occipital "driving" and average visual evoked potentials). *Electroencephalog. clin. Neurophysiol.* **23**: 383–384.

Price, L. L., and Goldstein, R. (1966). Averaged evoked responses for measuring auditory sensitivity in children. *J. Speech Hearing Disorders* **31** : 248–256.

Prieto, S., Villar, J. I., and Bachini, O. (1965). Influencias farmacologicas sobre la respuesta visual provocada en el hombre. *Acta Neurol. Latinoamer.* **11** : 295–296.

Rapin, I. (1964). Evoked responses to clicks in a group of children with communication disorders. *Ann. N. Y. Acad. Sci.* **112**: 182–203.

Raven, J. C. (1958). *Standard Progressive Matrices, Sets, A, B, C, D, and F,* H. K. Lewis and Company, Ltd., London.

Rémond, A. (1964). Level of organization of evoked responses in man. *Ann. N. Y. Acad. Sci.* **112**: 143–159.

Rémond, A. (1967). In: H. Gastaut, F. Bostem, R. Poiré, A. Waltregny, and H. Regis (eds.), *Les Activités Électriques Cérébrales Spontanées et Evoquées Chez L'homme,* Gauthier-Villars, Paris, pp. 17–26.

Rémond, A. (1968). The importance of topographic data in EEG phenomena and an electrical model to reproduce them. *Electroencephalog. clin. Neurophysiol.* Suppl. **27**: 29–49.

Rhodes, L. E., Dustman, R. E., and Beck, E. C. (1969). The visual evoked response: A comparison of bright and dull children. *Electroencephalog. clin. Neurophysiol.* **27**: 364–372.

Rietveld, W. J. (1963). The occipitocortical response to light flashes in man. *Acta Physiol. Pharmacol. Neerl.* **12**: 373–407.

Ritter, W., and Vaughan, H. G., Jr. (1969). Averaged evoked responses in vigilance and discrimination: A reassessment. *Science* **164**: 326–328.

Ritter, W., Vaughan, H. G., Jr., and Costa, L. D. (1968). Orienting and habituation to auditory stimuli: A study of short term changes in average evoked responses. *Electroencephalog. clin. Neurophysiol.* **25**: 550–556.

Rodin, E. A., and Luby, E. D. (1965). The effects of some psychosomimetic agents on visually evoked responses and background EEG. *Electroencephalog. clin. Neurophysiol.* **19**: 319.

Rodin, E., and Luby, E. (1966). Effects of LSD-25 on the EEG and photic evoked responses. *Arch. Gen. Psychiat.* **14**: 435–441.

Rodin, E., Zacharopoulos, G., Beckett, P., and Frohman, C. (1964). Characteristics of visually evoked responses in normal subjects and schizophrenic patients. *Electroencephalog. clin. Neurophysiol.* **17**: 458.

Rodin, E. A., Grisell, J. L., Gudobba, R. D., and Zachary, G. (1965). Relationship of EEG background rhythms to photic evoked responses. *Electroencephalog. clin. Neurophysiol.* **19**: 301–304.

Rodin, E., Grisell, J., and Gottlieb, J. (1968). Some electrographic differences between chronic schizophrenic patients and normal subjects. *In:* J. Wortis (ed.), *Recent Advances in Biological Psychiatry,* Vol. X, Plenum Press, New York, pp. 194–204.

Rosenfeld, J. P., Rudell, A. P., and Fox, S. S. (1969). Operant control of neural events in humans. *Science* **165**: 821–823.

Rosner, B. S., Allison, T., Swanson, E., and Goff, W. R. A. (1960). A new instrument for the summation of evoked responses from the nervous system. *Electroencephalog. clin. Neurophysiol.* **12**: 745–747.

Rosner, B. S., Goff, W. R., and Allison, T. (1963). Cerebral electrical responses to external stimuli. *In:* G. H. Glaser (ed.), *EEG and Behavior.* Basic Books, New York, pp. 109–133.

Rubin, L. S. (1962). Patterns of adrenergic–cholinergic imbalance in the functional psychoses. *Psychol. Rev.* **69**: 501–519.

Ruchkin, D. S., Villegas, J., and John, E. R. (1964). An analysis of average evoked potentials making use of least mean square techniques. *Ann N. Y. Acad. Sci.* **115**: 799–826.

Rynearson, R. R., Wilson, M. R., and Bickford, R. G. (1968). Psilocybin-induced changes in psychologic function, electroencephalogram, and light-evoked potentials in human subjects. *Mayo Clinic Proc.* **43**: 191–204.

Saier, J., Régis, H., Mano, T., and Gastaut, H. (1968). Potentiels évoqués visuels et somesthésiques pendant le sommeil de l'homme. *Brain Res.* **10**: 431–440.

Saletu, B., Saletu, M., Itil, T., and Jones, J. (1970). Somatosensory evoked potential changes during thiothixene treatment in schizophrenic patients. Presented at American EEG Society, Washington, D. C., Sept. 1970.

Satterfield, J. H. (1965). Evoked cortical response enhancement and attention in man. A study of responses to auditory and shock stimuli. *Electroencephalog. clin. Neurophysiol.* **19**: 470–475.

Satterfield, J. H. (1966). System for selection of responses for averaging. *Electroencephalog. clin. Neurophysiol.* **21**: 86–88.

Satterfield, J. H. (1969a). *Discussion in:* E. Donchin and D. B. Lindsley (eds.), *Average Evoked Potentials: Methods, Results, and Evaluations,* National Aeronautics and Space Administration, Washington, D. C., pp. 320–321.

Satterfield, J. H. (1969b). Auditory evoked cortical response studies in depressed patients and normal control subjects. Presented at NIMH Workshoo on the Psychobiology of the Depressive Illnesses, in press.

Satterfield, J.H., and Cheatum, D. (1964). Evoked cortical potential correlates of attention in human subjects. *Electroencephalog. clin. Neurophysiol.* **17**: 456.

Schulte, F.J., Akiyama, Y., and Parmalee, A.H. (1967). Auditory evoked responses during sleep in premature and full-term newborn infants. *Electroencephalog. clin. Neurophysiol.* **23**: 97.

Schwartz, M., and Shagass, C. (1961). Physiological limits for "subliminal" perception. *Science* **133**: 1017–1018.

Schwartz, M., and Shagass, C. (1963). Reticular modification of somatosensory cortical recovery function. *Electroencephalog. clin. Neurophysiol.* **15**: 265–271.

Schwartz, M., and Shagass, C. (1964). Recovery functions of human somatosensory and visual evoked potentials. *Ann. N. Y. Acad. Sci.* **112**: 510–525.

Schwartz, M., Emde, J., and Shagass, C. (1964). Comparison of constant current and constant voltage stimulators for scalp-recorded somatosensory responses. *Electroencephalog. clin. Neurophysiol.* **17**: 81–83.

Shagass, C. (1954). The sedation threshold. A method for estimating tension in psychiatric patients. *Electroencephalog. clin. Neurophysiol.* **6**: 221–223.

Shagass, C. (1955a). Differentiation between anxiety and depression by the photically activated electroencephalogram. *Amer. J. Psychiat.* **112**: 41–46.

Shagass, C. (1955b). Anxiety, depression and the photically driven electroencephalogram. *Arch. Neurol. Psychiat.* **74**: 3–10.

Shagass, C. (1967a). Effects of LSD on somatosensory and visual evoked responses and on the EEG in man. *In:* J. Wortis (ed.), *Recent Advances in Biological Psychiatry,* Vol. IX, Plenum Press, New York, pp. 209–227.

Shagass, C. (1967b). Discussion of EEG and Intelligence by W.T. Liberson. *In:* J. Zubin and G.A. Jervis (eds.), *Psychopathology of Mental Development,* Grune and Stratton, New York, pp. 626–628.

Shagass, C. (1967c). Sedation threshold: Technique and concept. *In:* H. Brill, J.O. Cole, P. Deniker, H. Hippius, and P.B. Bradley (eds.), *Neuropsychopharmacology,* Excerpta Medica Foundation, Amsterdam, pp. 921–925.

Shagass, C. (1968a). Averaged somatosensory evoked responses in various psychiatric disorders. *In:* J. Wortis (ed.), *Recent Advances in Biological Psychiatry,* Vol. X, Plenum Press, New York, pp. 205–219.

Shagass, C. (1968b). Evoked responses in psychiatry. *In:* N.S. Kline and E. Laska (eds.), *Computers and Electronic Devices in Psychiatry,* Grune and Stratton, New York, pp. 141–157.

Shagass, C. (1968c). Pharmacology of evoked potentials in man. *In:* D.H. Efron (ed.), J.O. Cole, J. Levine, and J.R. Wittenborn (co-eds.), *Psychopharmacology: A Review of Progress 1957–1967,* Public Health Service Publication No. 1836, pp. 483–492.

Shagass, C. (1968d). Cerebral evoked response findings in schizophrenia, *Cond. Reflex* **3**: 205–216.

Shagass, C. (1968e). Evoked response and behavioral effects of LSD and Ditran. *In:* D.H. Efron (ed.), J.O. Cole, J. Levine, and J.R. Wittenborn (co-eds.), *Psychopharmacology: A Review of Progress 1957–1967,* Public Health Service Publication No. 1836, pp. 1241–1246.

Shagass, C., and Ando, K. (1970). Septal and reticular influences on cortical evoked response recovery functions. *Biol. Psychiat.* **2**: 3–18.

Shagass, C., and Bittle, R.M. (1967). Therapeutic effects of LSD: A follow-up study. *J. Nerv. Ment. Dis.* **144**: 471–478.

Shagass, C., and Canter, A. (1966). Some personality correlates of cerebral evoked response characteristics. *Proc. XVIII Int. Congr. Psycho.,* Symposium No. 6, Moscow, pp. 47–52.

Shagass, C., and Overton, D.A. (1970). Measurement of cerebral "excitability characteristics" in relation to psychopathology. *In:* M. Kietzman and J. Zubin (eds.), *Objective Indicators of Psychopathology,* Academic Press, New York (in press).

Shagass, C., and Schwartz, M. (1961a). Evoked cortical potentials and sensation in man. *J. Neuropsychiat.* **2**: 262–270.

Shagass, C., and Schwartz, M. (1961*b*). Reactivity cycle of somatosensory cortex in humans with and without psychiatric disorder. *Science* **134**: 1757–1759.

Shagass, C., and Schwartz, M. (1961*c*). Cortical excitability in psychiatric disorder. Preliminary results. *Proc. III World Congress Psychiatry*, Vol. 1, University of Toronto Press, Toronto, pp. 441–446.

Shagass, C., and Schwartz, M. (1962*a*). Excitability of the cerebral cortex in psychiatric disorder. *In:* R. Roessler and N.S. Greenfield (eds.), *Physiological Correlates of Psychological Disorder*, University of Wisconsin Press, Madison, pp. 45–60.

Shagass, C., and Schwartz, M. (1962*b*). Observations on somatosensory cortical reactivity in personality disorder. *J. Nerv. Ment. Dis.* **135**: 44–51.

Shagass, C., and Schwartz, M. (1962*c*). Cerebral cortical reactivity in psychotic depressions. *Arch. Gen. Psychiat.* **6**: 235–242.

Shagass, C., and Schwartz, M. (1962*d*). Effects of stimulus intensity on the form and recovery cycle of evoked somatosensory cortical potentials in man. *Proc. XXII Internat. Congress of Physiological Sciences*, No. 1056.

Shagass, C., and Schwartz, M. (1963*a*). Cerebral responsiveness in psychiatric patients. *Arch. Gen. Psychiat.* **8**: 177–189.

Shagass, C., and Schwartz, M. (1963*b*). Psychiatric disorder and deviant cerebral responsiveness to sensory stimulation. *In:* J. Wortis (ed.), *Recent Advances in Biological Psychiatry*. Volume V, Plenum Press, New York, pp. 321–330.

Shagass, C., and Schwartz, M. (1963*c*). Psychiatric correlates of evoked cerebral cortical potentials. *Amer. J. Psychiat.* **119**: 1055–1061.

Shagass, C., and Schwartz, M. (1964*a*). Recovery functions of somatosensory peripheral nerve and cerebral evoked responses in man. *Electroencephalog. clin. Neurophysiol.* **17**: 126–135.

Shagass, C., and Schwartz, M. (1964*b*). Evoked potential studies in psychiatric patients. *Ann. N. Y. Acad. Sci.* **112**: 526–542.

Shagass, C., and Schwartz, M. (1965*a*). Visual cerebral evoked response characteristics in a psychiatric population. *Amer. J. Psychiat.* **121**: 979–987.

Shagass, C., and Schwartz, M. (1965*b*). Age, personality and somatosensory evoked responses. *Science* **148**: 1359–1361.

Shagass, C., and Schwartz, M. (1966). Somatosensory cerebral evoked responses in psychotic depression. *Brit. J. Psychiat.* **112**: 799–807.

Shagass, C., and Trusty, D. (1966). Somatosensory and visual cerebral evoked response changes during sleep. *In:* J. Wortis (ed.) *Recent Advances in Biological Psychiatry*, Vol. VIII, Plenum Press, New York, pp. 321–334.

Shagass, C., Schwartz, M., and Amadeo, M. (1962). Some drug effects on evoked cerebral potentials in man. *J. Neuropsychiat.* **3**: S49–S58.

Shagass, C., Schwartz, M., and Krishnamoorti, S.R. (1965). Some psychologic correlates of cerebral responses evoked by light flash. *J. Psychosom. Res.* **9**: 223–231.

Shagass, C., Häseth, K., Callaway, E., and Jones, R.T. (1968). EEG-evoked response relationships and perceptual performance. *Life Sci.* **19**: 1083–1091.

Shagass, C., Overton, D.A., and Bartolucci, G. (1969). Evoked responses in schizophrenia. *In:* D.V. Siva Sankar (ed.), *Schizophrenia: Current Concepts and Research*, PJD Publications, Ltd., Hicksville, New York, pp. 220–235.

Shagass, C., Overton, D.A., Bartolucci, G., and Straumanis, J.J. (1970). Effect of attention modification by television viewing on somatosensory evoked responses and recovery function. *J. Nerv. Ment. Dis.,* in press.

Shagass, C., Overton, D.A., and Straumanis, J.J. (1971). Evoked response and EEG findings in psychiatric illness related to drug abuse. *Biol. Psychiat.,* in press.

Shakow, D. (1963). Psychological deficit in schizophrenia. *Behav. Sci.* **8**: 275–305.

Shevrin, H., and Fritzler, D.E. (1968). Visual evoked response correlates of unconscious mental processes. *Science* **161**: 295–298.

Shevrin, H., and Rennick, P. (1967). Cortical response to a tactile stimulus during attention, mental arithmetic, and free associating. *Psychophysiol.* **3**: 381–388.

Shevrin, H., Smith, W. H., and Fritzler, D. E. (1969). Repressiveness as a factor in the subliminal activation of brain and verbal responses. *J. Nerv. Ment. Dis.* **140**: 261–269.

Shipley, T., Jones, R. W., and Fry, A. (1965). Evoked visual potentials and human color vision. *Science* **150**: 1162–1164.

Shipton, H. W. (1960). A photographic averaging technique for the study of evoked cortical potentials in man. *In:* C. N. Smyth (ed.), *Medical Electronics,* Iliffe, London, pp. 186–187.

Short, M. J., Wilson, W. P., and Gills, J. P., Jr. (1966). Thyroid hormone and brain function. IV. Effect of triiodothyronine on visual evoked potentials and electroretinogram in man. *Electroencephalog. clin. Neurophysiol.* **25**: 123–127.

Shucard, D. W. (1969). Relationships among measures of the cortical evoked potential and abilities comprising human intelligence, Ph.D. thesis, University of Denver.

Silverman, J. (1967). Variations in cognitive control and psychophysiological defense in the schizophrenias. *Psychosom. Med.* **29**: 225–251.

Small, J. G., and Small, I. F. (1969). Expectancy wave in affective psychoses. Presented at NIMH Workshop on *Recent Advances in the Psychology of the Depressive Illnesses,* Williamsburg, Virginia, (May, 1969).

Small, J. G., Small, I. F., and Perez, H. C. (1970). Evoked and slow potential variations with lithium in manic-depressive disease. Presented at Society of Biological Psychiatry, San Francisco, California (May 1970).

Speck, L. B., Dim, B., and Mercer, M. (1966). Visual evoked responses of psychiatric patients. *Arch. Gen. Psychiat.* **15**: 59–63.

Spehlmann, R. (1965). The averaged electrical responses to diffuse and to patterned light in the human. *Electroencephalog. clin. Neurophysiol.* **19**: 560–569.

Spilker, B., and Callaway, E. (1969a). "Augmenting" and "reducing" in averaged visual evoked responses to sine wave light. *Psychophysiology* **6**:49–57.

Spilker, B., and Callaway, E. (1969b). Effects of drugs on "augmenting/reducing" in averaged visual evoked responses in man. *Psychopharmacologia* **15**: 116–124.

Spilker, B., Kamiya, J., Callaway, E., and Yeager, C. L. (1969). Visual evoked responses in subjects trained to control alpha rhythms. *Psychophysiology* **5**: 683–695.

Spong, P., Haider, M., and Lindsley, D. B. (1965). Selective attentiveness and cortical evoked responses to visual and auditory stimuli. *Science* **148**: 395–397.

Straumanis, J., Shagass, C., and Schwartz, M. (1965). Visually evoked cerebral response changes associated with chronic brain syndrome and aging. *J. Gerontol.* **20**: 498–506.

Straumanis, J., Shagass, C., and Overton, D. A. (1969a). Problems associated with application of the contingent negative variation to psychiatric research. *J. Nerv. Ment. Dis.* **148**: 170–179.

Straumanis, J., Shagass, C., and Overton, D. A. (1969b). Evoked responses in Down's syndrome of young adults. Presented at Eastern Association of EEG meeting, New York (December 1969).

Sutton, S. (1969). The specification of psychological variables in an average evoked potential experiment. *In:* E. Donchin and D. Lindsley (eds.), *Average Evoked Potentials: Methods, Results, and Evaluations,* National Aeronautics and Space Administration, Washington, D. C., pp. 237–297.

Sutton, S., Braren, M., and Zubin, J. (1965). Evoked-potential correlates of stimulus uncertainty. *Science* **150**: 1187–1188.

Sutton, S., Tueting, P., Zubin, J., and John, E. R. (1967). Information delivery and the sensory evoked potential. *Science* **155**: 1436–1439.

Suzuki, T., and Taguchi, K. (1968). Cerebral evoked response to auditory stimuli in young children during sleep. *Ann. Otol. Rhinol. Laryng.* **77**: 102–110.

Tecce, J. J. (1970). Attention and evoked potentials in man. *In:* D. I. Mostofsky (ed.), *Attention: Contemporary Theory and Analysis,* Appleton-Century-Crofts, New York, pp. 331–365.

Tecce, J. J., and Scheff, N. M. (1969). Attention reduction and suppressed direct-current potentials in the human brain. *Science* **164**: 331–333.

Tepas, D. I., Armington, J. C., and Kropel, W. J. (1962). Evoked potentials in the human visual system. *In:* E. E. Bernard and M. R. Kare (eds.), *Biological Prototypes and Synthetic Systems,* Volume 1, Plenum Press, New York, pp. 13–21.

Teuting, P. A. (1968). Uncertainty and averaged evoked response in a guessing situation, Doctoral Thesis, Columbia University.

Thurstone, L. L. (1944). *A Factorial Study of Perception,* University of Chicago Press, Chicago.

Thurstone, L. L., and Jeffrey, T. E. (1965). *Closure Flexibility (Concealed Figures) (Form A), Test Administration Manual,* Industrial Relations Center, University of Chicago, Chicago.

Thurstone, L. L., and Jeffrey, T. E. (1966). *Closure Speed, Test Administration Manual,* Industrial Relations Center, University of Chicago, Chicago.

Timsit, M., Koninckx, N., Dargent, J., Fontaine, O., and Dongier, M. (1969). Étude de la durée des VCN chez un groupe de sujets normaux un groupe de névrosés et un groupe de psychotiques. *In:* J. Dargent and M. Dongier (eds.), *Variations Contingentes Négatives.* University of Liège, Belgium, pp. 206–214.

Ulett, G. A., Gleser, G., Winokur, G., and Lawler, A. (1953). The EEG and reaction to photic stimulation as an index of anxiety-proneness. *Electroencephalog. clin. Neurophysiol.* **5**: 23–32.

Ulett, G. A., Heusler, A. F., and Word, T. J. (1965). The effect of psychotropic drugs on the EEG of the chronic psychotic patient. *In:* W. P. Wilson (ed.), *Applications of Electroenophalography to Psychiatry,* Duke University Press, Durham, pp. 241–257.

Utall, W. R. (1965). Do compound evoked potentials reflect psychological codes? *Psychol. Bull.* **64**: 377–392.

Utall, W. R., and Cook, L. (1964). Systematics of the evoked somatosensory cortical potential: A psychophysical-electrophysiological comparison. *Ann. N. Y. Acad. Sci.* **112**: 60–80.

Van Balen, A. T. M. (1960). *De elektroencefalografische Reactie op Lichtprikkeling en zijn Betekenis voor de oogheelkundige Diagnostiek,* Doctoral Thesis, University of Utrecht, Klimp en Bowman, Rotterdam.

Van Der Tweel, L. H., and Verduyn-Lunel, H. F. E. (1965). Human visual responses to sinusoidally modulated light. *Electroencephalog. clin. Neurophysiol.* **18**: 587–598.

Van Hof, W. M. (1960). The relation between the cortical responses to flash and to flicker in man. *Acta Physiol. Pharmacol. Neerl.* **9**: 210–224.

Van Meter, W. G., Owens, H. F., and Himwich, H. E. (1959). Effects of tofranil, an antidepressant drug, on electrical potentials of rabbit brain. *Canad. Psychiat. Ass. J.* 4:S113–S119.

Vanzulli, A., Bogacz, J., Handler, P., and Garcia Austt, E. (1960). Evoked responses in man. I. Photic responses. *Acta Neurol. Latinomer.* **6**: 219–231.

Vanzulli, A., Bogacz, J., and Garcia-Austt, E. (1961). Evoked responses in man. III. Auditory response. *Acta. Neurol. Latinoamer.* **7**: 303–309.

Vaughan, H. G. (1969). The relationship of brain activity to scalp recordings of event-related potentials. *In:* E. Donchin and D. B. Lindsley (eds.), *Average Evoked Potentials: Methods, Results, and Evaluations,* National Aeronautics and Space Administration, Washington, D. C., pp. 45–75.

Vaughan, H. G., Costa, L. D., and Ritter, W. (1968). Topography of the human motor potential. *Electroencephalog. clin. Neurophysiol.* **25**: 1–10.

Venables, P. H. (1963). The relationship between level of skin potential and fusion of paired light flashes in schizophrenic and normal subjects. *J. Psychiat. Res.* **1**: 279–287.

Walter, W. G. (1936). The location of cerebral tumors by electroencephalography. *Lancet* **2**: 305–308.

Walter, W. G. (1962). Spontaneous oscillatory systems and alterations in stability. *In:* R. G. Grenell (ed.), *Neural Physiopathology,* Hoeber, New York, pp. 222–257.

Walter, W. G. (1964*a*). The convergence and interaction of visual, auditory, and tactile responses in human nonspecific cortex. *Ann. N. Y. Acad. Sci.* **112**: 320–361.

Walter, W. G. (1964*b*). Slow potential waves in the human brain associated with expectancy, attention and decision. *Arch. Psychiat. Nervenkr.* **206**: 309–322.

Walter, W. G. (1966). Electrophysiologic contributions to psychiatric therapy. *In:* Current *Psychiatric Therapies,* Vol. VI, Grune and Stratton, New York, pp. 13–25.

Walter, W. G. (1967). Slow potential changes in the human brain associated with expectancy, decision and intention. *Electroencephalog. clin. Neurophysiol.* Suppl. **26**:123–130.

Walter, W. G. (1969). *In:* J. Dargent and M. Dongier (eds.), *Variations Contingentes Négatives,* University of Liège, Belgium, pp. 57–58, 76–77, 95–96.

Walter, W. G. (1970). The contingent negative variation as an aid to psychiatric diagnosis. *In:* M. Kietzman and J. Zubin (eds.), *Objective Indicators of Psychopathology,* Academic Press, New York, in press.

Walter, W. G., Cooper, R., Aldridge, V. J., McCallum, W. C., and Winter, A. L. (1964). Contingent negative variation: An electric sign of sensori-motor association and expectancy in the human brain. *Nature* **203**: 380–384.

Weinmann, H., Creutzfeldt, O., and Heyde, G. (1965). The development of the visual evoked response in children. *Arch. Psychiat. Nervenkr.* **207**: 323–341.

Weitzman, E. D. and Kremen, H. (1965). Auditory evoked responses during different stages of sleep in man. *Electroencephalog. clin. Neurophysiol.* **18**: 65–70.

Weitzman, E. D., Graziani, L., and Durhamel, L. (1967). Maturation and topography of the auditory evoked response of the prematurely born infant. *Electroencephalog. clin. Neurophysiol.* **23**: 82–83.

Welsh, G. S. (1965). MMPI profiles and factor scales A and R. *J. Consult. Psychol.* **21**: 43–47.

Werre, P. F., and Smith, C. J. (1964). Variability of responses evoked by flashes in man. *Electroencephalog. clin. Neurophysiol.* **17**: 644–652.

Whitaker, H. S., Osborne, R. T., and Nicora, B. (1967). Intelligence measured by analysis of the photic evoked response. Paper presented to American Neurological Assn. (June 14, 1967).

Wilkinson, R. T., and Morlock, H. C. (1967). Auditory evoked responses and reaction time. *Electroencephalog. clin. Neurophysiol.* **23**: 50–56.

Williams, H. L., Tepas, D. I., and Morlock, H. C. (1962). Evoked responses to clicks and electroencephalographic stages of sleep in man. *Science* **138**: 685–686.

Williams, H. L., Morlock, H. C., Jr., Morlock, J. V., and Lubin, A. (1964). Auditory evoked responses and the EEG stages of sleep. *Ann. N. Y. Acad. Sci.* **112**: 172–181.

Wilson, R., and Shagass, C. (1964). Comparison of two drugs with psychotomimetic effects (LSD and Ditran). *J. Nerv. Ment. Dis.* **138**: 277–286.

Winters, W. D. (1964). Comparison of the average cortical and subcortical evoked response to clicks during various stages of wakefulness, slow wave sleep and rhombencephalic sleep. *Electroencephalog. clin. Neurophysiol.* **17**:234–245.

Witkin, H. A., Lewis, H. B., Hertzman, M., Machover, K., Meissner, P. B., and Wapner, S. (1954). *Personality Through Perception,* Harper and Row, New York, p. 571.

Woodworth, R. S., and Schlosberg, H. (1954). *Experimental Psychology,* Holt, Rinehart and Winston, New York, pp. 200–233.

Index